Fundamentals of Vehicle Dynamics and Modelling

Automotive Series

Advanced Battery Management Technologies for Electric Vehicles
Rui Xiong, Weixiang Shen

Noise and Vibration Control in Automotive Bodies
Jian Pang

Automotive Power Transmission Systems
Yi Zhang, Chris Mi

High Speed Off-Road Vehicles: Suspensions, Tracks, Wheels and Dynamics
Bruce Maclaurin

Hybrid Electric Vehicles: Principles and Applications with Practical Perspectives, 2nd Edition
Chris Mi, M. Abul Masrur

Hybrid Electric Vehicle System Modeling and Control, 2nd Edition
Wei Liu

Thermal Management of Electric Vehicle Battery Systems
Ibrahim Dincer, Halil S. Hamut, Nader Javani

Automotive Aerodynamics
Joseph Katz

The Global Automotive Industry
Paul Nieuwenhuis, Peter Wells

Vehicle Dynamics
Martin Meywerk

Modelling, Simulation and Control of Two-Wheeled Vehicles
Mara Tanelli, Matteo Corno, Sergio Saveresi

Vehicle Gearbox Noise and Vibration: Measurement, Signal Analysis, Signal Processing and Noise Reduction Measures
Jiri Tuma

Modeling and Control of Engines and Drivelines
Lars Eriksson, Lars Nielsen

Advanced Composite Materials for Automotive Applications: Structural Integrity and Crashworthiness
Ahmed Elmarakbi

Guide to Load Analysis for Durability in Vehicle Engineering
P. Johannesson, M. Speckert

Fundamentals of Vehicle Dynamics and Modelling

A Textbook for Engineers with Illustrations and Examples

Bruce P. Minaker
University of Windsor
ON, CA

WILEY

This edition first published 2019

Registered Offices
John Wiley & Sons, Inc., 111 River Street, Hoboken, NJ 07030, USA
John Wiley & Sons Ltd, The Atrium, Southern Gate, Chichester, West Sussex, PO19 8SQ, UK

Editorial Office
The Atrium, Southern Gate, Chichester, West Sussex, PO19 8SQ, UK

For details of our global editorial offices, customer services, and more information about Wiley products visit us at www.wiley.com.

Wiley also publishes its books in a variety of electronic formats and by print-on-demand. Some content that appears in standard print versions of this book may not be available in other formats.

Library of Congress Cataloging-in-Publication Data

Names: Minaker, Bruce P., 1969- author.
Title: Fundamentals of vehicle dynamics and modelling : a textbook for engineers with illustrations and examples / Bruce P. Minaker, University of Windsor, ON, CA.
Description: Hoboken, NJ : Wiley, 2020. | Series: Automotive series | Includes bibliographical references and index. |
Identifiers: LCCN 2019020054 (print) | LCCN 2019021720 (ebook) | ISBN 9781118980071 (Adobe PDF) | ISBN 9781118980088 (ePub) | ISBN 9781118980095 (hardcover)
Subjects: LCSH: Motor vehicles–Dynamics–Textbooks. | Motor vehicles–Mathematical models–Textbooks.
Classification: LCC TL243 (ebook) | LCC TL243 .M555 2020 (print) | DDC 629.2/31–dc23
LC record available at https://lccn.loc.gov/2019020054

Cover Design: Wiley
Cover Images: Background: © solarseven/Shutterstock, Left: © Denis Vorob'yev/iStockphoto, Middle: © Henrik5000/Getty Images, Right: © zorazhuang/Getty Images

Set in 10/12pt WarnockPro by SPi Global, Chennai, India

Printed and bound by CPI Group (UK) Ltd, Croydon, CR0 4YY

10 9 8 7 6 5 4 3 2 1

To my wife Beth Anne, for everything

Contents

List of Figures

List of Tables

Preface

This is my first attempt at writing a book. After many years of effort, I realize now that I was unprepared for such a task when I began, but I did it in the hopes that some of you may find it useful. What prompted this foray into the textbook market? Over many years of university teaching, I have found many texts on this topic and others that, while providing good relevant coverage, add much more content than necessary. I believe that this level of breadth is not in the reader's best interest, with many left overwhelmed by the prospect of mastering so much material, and eventually resigning themselves to never finishing. It is intimidating for a student to contemplate absorbing over a thousand pages in a twelve-week semester.

My objective was to produce a book that is suitable for a single-semester senior undergraduate or early graduate level mechanical engineering course, perhaps in a program with some focus on automotive engineering. It assumes that the student has some foundation in mathematics, particularly linear algebra and differential equations, and basic rigid body dynamics. The book does not aim to be a complete reference, but rather to give a solid foundation while generating enthusiasm in the student reader. I have found vehicle dynamics to be a most intellectually rewarding topic, and if I can share some of that with my readers, then I have accomplished my goal.

The book opens with some brief general discussion of the topic, and follows with an introduction to tire modelling. It includes material on longitudinal dynamics, lateral dynamics, and vertical dynamics. I have included many of the classic models that first drew me in, problems small enough to solve by hand on paper, without the use of a computer. This leads into some larger problems, tied to my work on automatic generation of the equations of motion of multibody systems, and its application to vehicle dynamics. The text concludes with a chapter expanding the necessary mathematical background for those readers who may require it.

I would like the reader to come away with answers to the three following questions: First, what are the concepts and tools used to generate a mathematical model of vehicle motion? Next, what are the fundamental vehicle ride and handling behaviours that we can predict from well-established models? Finally, how has the arrival of multibody dynamics and computer aided engineering changed what we can do now? Hopefully, this will remove some of the mystery behind many of the software tools used in industry today, and how they generate at their results.

Astute readers may notice what seems like an odd combination of British and American English styles in the text. This stems from my education in and use of Canadian English. You can expect '-our' rather than '-or', but '-ize' instead of '-ise'. In any instance where there was flexibililty in spelling, punctuation or grammar, I have tried to emulate those pre-war vintage textbooks that I enjoy collecting.

Bruce P. Minaker
Windsor, Ontario, Canada

List of Symbols

A mathematical style consistent with formatting standards is used wherever possible throughout the text. This has proven to be surprisingly difficult, as the symbols that have typically been associated with certain quantities tend to be inconsistent between fields, or overlapping. As a result, a few concessions and stylistic choices have been made, e.g., the character **L** has been used to indicate a damping matrix, in place of the more commonly used **C**, as the latter is also used in the standard state space form. The lowercase c is still used to indicate a damping coefficient in the scalar case.

Upright characters indicate a mathematical constant ($\mathrm{\pi}$, e, $\mathrm{i} = \sqrt{-1}$), italics are used to indicate a quantity is a mathematical variable, (x, ω), and boldface is used to indicate a vector quantity, ($\boldsymbol{x}$, $\boldsymbol{\omega}$). Lowercase characters are chosen wherever possible, with the exception of matrices, which are always set in bold upright uppercase ($\mathbf{M}$, $\mathbf{K}$). When the appropriate lowercase character is already in use, an uppercase character may be substituted. Historically, certain quantities (e.g., scalar components of force and moment vectors) have been set in uppercase; these choices are maintained when appropriate. No distinction is made between Latin and Greek characters when formatting; the traditional Greek characters are maintained for several quantities. Multicharacter names are avoided, to eliminate confusion with products of variables. Note that these rules are also applied to subscripts, i.e., if a quantity is marked with an italic subscript, that subscript is itself a variable. An upright subscript indicates a specific named instance of the variable. Any character modified with a dot above indicates the time derivative of that quantity ($\dot{x}$, $\dot{\boldsymbol{\omega}}$). A vector modified with a circumflex indicates a unit vector ($\hat{\boldsymbol{u}}$). The standard basis vectors are $\hat{\mathbf{\imath}}$, $\hat{\mathbf{\jmath}}$, and $\hat{\mathbf{k}}$. A vector modified with a tilde indicates the skew symmetric matrix of the vector ($\tilde{\boldsymbol{r}}$). A pair of vertical bars around a character is used to indicate the absolute value of a scalar, the magnitude of a vector or a complex number, or the amplitude of a time varying sinusoid ($|x|$, $|\boldsymbol{x}|$). The transpose of a matrix or a vector is indicated using a prime ($\boldsymbol{x}'$). Column vectors may be presented as a row by using transpose notation for the space savings offered.

In 1976, the Society of Automotive Engineers (SAE) published standard J670e, establishing a convention for terminology and notation for vehicle dynamics. In 1991, the International Organization for Standardization (ISO) published a vehicle dynamics vocabulary, ISO 8855. The SAE J670e and ISO 8855 standards are incompatible in several aspects, the most notable being the axis systems defined in the two documents. The SAE standard utilizes an axis system based on aeronautical practice, with the x axis positive forward, the y axis positive to the driver's right, and the z axis positive down. The ISO standard utilizes an axis system with the x axis positive forward, the y axis positive to the driver's left, and the z axis positive up. In this text, the SAE standard and the corresponding historical aeronautical notation will be used in the discussion of the vehicle models in Chapters 3, 4, and 5, as they were developed well before 1991. In Chapter 6, an ISO-style axis system will be used when discussing a modern multibody dynamics approach.

Scalar symbols

Symbol	Description
a	longitudinal distance from mass centre to front axle
b	longitudinal distance from mass centre to rear axle
c	damping coefficient, or tire cornering coefficient
c_d	drag coefficient
c_n	arbitrary constant coefficient
d_n	arbitrary constant coefficient
d	longitudinal distance from mass centre to trailer hitch
e	longitudinal distance from trailer mass centre to trailer hitch
f	force, or arbitrary function
g	gravitational acceleration
h	longitudinal distance from trailer mass centre to trailer axle, or time step
h_G	centre of mass height
i	counter increment
k	spring stiffness

Symbol	Description
k_{yy}	pitch radius of gyration
k_{zz}	yaw radius of gyration
l	length
m	mass
n	counter end, or dimension
p	angular velocity, x axis direction
q	angular velocity, y axis direction
r	angular velocity, z axis direction
s	eigenvalue, or exponent coefficient
t	time, or track width
u	linear velocity, x axis direction
v	linear velocity, y axis direction
w	linear velocity, z axis direction
x	location, x axis direction
y	location, y axis direction
z	location, z axis direction
G	centre of mass
I_{xx}	roll moment of inertia
I_{yy}	pitch moment of inertia
I_{zz}	yaw moment of inertia
A_f	frontal area
L	moment, x axis direction
M	moment, y axis direction
N	moment, z axis direction
P	power
R	cornering radius
T	time step size, discrete time
X	force, x axis direction
Y	force, y axis direction
Z	force, z axis direction

Vector or matrix symbols

Symbol	Description
$\boldsymbol{a}$	acceleration vector
$\boldsymbol{f} = X\hat{\mathbf{i}} + Y\hat{\mathbf{j}} + Z\hat{\mathbf{k}}$	force vector
$\boldsymbol{m} = L\hat{\mathbf{i}} + M\hat{\mathbf{j}} + N\hat{\mathbf{k}}$	moment vector
$\boldsymbol{p}$	location and orientation vector
$\boldsymbol{r}$	radius vector
$\boldsymbol{s}$	spin vector
$\boldsymbol{u}$	input signal
$\hat{\boldsymbol{u}}$	unit vector
$\boldsymbol{w}$	linear and angular velocity vector
$\boldsymbol{x}$	state vector, location vector, eigenvector
$\boldsymbol{y}$	output vector
$\boldsymbol{z}$	vertical location vector
$\mathbf{A}$	system matrix
$\mathbf{B}$	input matrix
$\mathbf{C}$	output matrix
$\mathbf{D}$	feedthrough matrix
$\mathbf{E}$	descriptor matrix
$\mathbf{F}$	input force matrix
$\mathbf{G}$	input rate force matrix, or transfer function matrix
$\mathbf{H}$	deflection Jacobian matrix
$\mathbf{I}$	identity matrix
$\mathbf{I}_{\mathrm{G}}$	inertia matrix
$\mathbf{J}$	constraint Jacobian matrix
$\mathbf{K}$	stiffness matrix
$\mathbf{L}$	damping matrix
$\mathbf{M}$	mass matrix
$\mathbf{P}$	transformation matrix
$\mathbf{R}$	rotation matrix
$\mathbf{S}$	angular velocity transformation matrix
$\mathbf{T}$	orthogonal complement matrix
$\mathbf{U}$	orthogonal complement matrix
$\mathbf{V}$	velocity matrix

Greek symbols

Symbol	Description
α	tire slip angle
β	body slip angle
γ	trailer sway angle
δ	steer angle
ε	camber angle
ζ	damping ratio
η	efficiency
θ	pitch angle
κ	friction gradient
λ	wavelength
μ	coefficient of friction
ξ	jacking force angle
ρ	density
σ	real component of eigenvalue, or tire slip ratio
τ	time constant
ϕ	roll angle, or phase angle
ψ	yaw angle
ω	imaginary component of eigenvalue, or angular frequency
$\boldsymbol{\alpha}$	angular acceleration vector
$\boldsymbol{\Delta}$	elastic deflection vector
$\boldsymbol{\theta}$	angular position vector
$\boldsymbol{\lambda}$	Lagrange multiplier
$\mathbf{v} = u\hat{\mathbf{i}} + v\hat{\mathbf{j}} + w\hat{\mathbf{k}}$	linear velocity vector
$\boldsymbol{\phi}$	constraint equation vector
$\boldsymbol{\omega} = p\hat{\mathbf{i}} + q\hat{\mathbf{j}} + r\hat{\mathbf{k}}$	angular velocity vector

About the Companion Website

This book is accompanied by a companion website:

www.wiley.com/go/minaker/vehicle-dynamics

The website includes:

- Matlab codes

Scan this QR code to visit the companion website.

1

Introduction

The subject of vehicle dynamics is very broad, but it can be loosely defined as an analysis of the motion of passenger vehicles, with an intent to characterize the various behaviours, and understand their causes. In this text, the focus will be on four-wheeled road vehicles, but the topic is often expanded to include the study of multi-axle trucks and trailers, bicycles and motorcycles, rail vehicles, and even boats and aircraft, as many of the principles used are the same. If the focus is restricted to road vehicles, the topic can be broken into sections:

Tires The development of mathematical models to predict tire forces, often based on empirical or semi-empirical techniques, i.e., not derived from first principles, but rather measurements of tire properties and data from experiments

Longitudinal dynamics Acceleration and braking capability, coupling between engine and vehicle through the drivetrain, and weight transfer effects on tire traction

Ride quality A study of vertical dynamics and the ability of the vehicle's suspension to accommodate varying terrain while maintaining passenger comfort

Handling The behaviour of the vehicle with respect to its motion in the plane of the road, particularly its directional stability and its response to steering inputs

Suspension kinematics The geometry of the suspension and steering components, and its effects on vehicle motion; sometimes expanded to include *elastokinematics*, the small changes in the geometry due to deflection under load

This text will attempt to introduce the reader to each of these topics, but they are organized slightly differently than in the list above. Chapter 2 will introduce the topic of tire behaviour, and Chapter 3 will cover longitudinal dynamics. Chapter 4 will discuss both ride quality and handling, and focus on the application of linear dynamic modelling to these problems. Chapter 5 expands the models to cover the full vehicle, and discusses the effects of suspension

Fundamentals of Vehicle Dynamics and Modelling: A Textbook for Engineers with Illustrations and Examples, First Edition. Bruce P. Minaker.

Companion website: www.wiley.com/go/minaker/vehicle-dynamics

geometry, and the solution of the suspension kinematics problem. Chapter 6 focuses on the application of multibody dynamics to vehicle problems, and how the previously developed models can be expanded using computer based techniques. Finally, Chapter 7 is an overview of the mathematics that is required for the many analyses in the text. The current chapter provides a brief list of some of the notable contributors to the field, and a general overview of the topic of vehicle dynamics as a whole.

1.1 Past, Present, and Future

The study of vehicle dynamics has developed along essentially the same timeline as the automobile itself, with the first papers on the subject appearing in the early 1900s, and significant contributions beginning around the 1930s. There have been a number of significant researchers in vehicle dynamics. Without a doubt, the author's work has been influenced by these pioneers, and any similarity in notation or technique is no coincidence. A few of them, along with a brief description of their work, are mentioned below.

Maurice Olley An outstanding engineer, his numerous accomplishments during his work for Rolls-Royce, General Motors and others are detailed in the book *Chassis Design: Principles and Analysis Based on Maurice Olley's Technical Notes* [11]. The results of his experiments in ride quality are still cited as the Olley 'Flat Ride' guidelines. Credited with the introduction of independent suspension in American cars, he was twice awarded the Crompton Medal, the highest award of the Institution of Automobile Engineers.

Leonard Segel In 1956, he published the landmark paper 'Theoretical prediction and experimental substantiation of the responses of the automobile to steering control' [2]. His work was instrumental in understanding the motion and behaviour of road vehicles.

J.R. Ellis Author of the classic 1969 text *Vehicle Dynamics* [3], and its follow-up *Vehicle Handling Dynamics* [9] in 1994.

Robin Sharp In addition to his numerous papers on automobile dynamics, he was a pioneer on the study of motorcycle dynamics. He wrote the 1971 paper 'The stability and control of motorcycles' [4].

David Crolla An expert in both road and off-road vehicle dynamics, his publication record includes some two hundred journal and conference papers. He served on the editorial boards of several journals, including *Proceedings of the IMechE Part D, Heavy Vehicle Systems,* and the *International Journal of Automotive Technology.*

Carol Smith Well known for his extensive hands-on experience and love of motorsport. He wrote the widely acclaimed *Tune to Win* [5] series of handbooks, which have motivated thousands of motor racing enthusiasts to study vehicle dynamics and engineering.

JY Wong An expert in terramechanics and off-road vehicles, and author of the widely cited text *Theory of Ground Vehicles* [6].

Hans Pacejka Developer of the now widely used 'magic' tire model, introduced in the 1987 paper 'Tire modelling for use in vehicle dynamics studies' [7], and further expanded in his text *Tire and Vehicle Dynamics* [12].

T.D. Gillespie The author of the text *Fundamentals of Vehicle Dynamics* [8], which was published by SAE and was a staple in the US auto industry for many years.

Bill Milliken The developer of the Milliken Moment Method, some say he invented the science of automobile handling. He's the co-author of the widely used text *Race Car Vehicle Dynamics* [10], and his own autobiography is given in *Equations of Motion: Adventure, Risk and Innovation* [13].

At this point, the author would be remiss not to make any mention of Ron Anderson, professor of Mechanical Engineering at Queen's University in Kingston, Ontario. While perhaps not as widely recognized as those listed above, his A'GEM (Automatic Generation of the Equations of Motion) software was instrumental in initiating the author's own research interest in the application of multibody dynamics techniques to the study of vehicle motion.

A review of the material produced by the authors listed above shows that initially the focus was on developing mathematical models to predict vehicle motion, and that those models were small enough that they could be manipulated on paper to determine the fundamental characteristics; most had from two to six degrees of freedom. As time passed, the models became more sophisticated, increasing in size and incorporating nonlinearities. The arrival of consumer-grade digital computing has had a profound effect, as multibody dynamics (MBD) has been applied to the vehicle problem in a number of ways. In the last decade or two, with the widespread availability of commercial MBD tools, the use of vehicle dynamics models in the auto industry has broadened considerably.

The ways in which modern automakers are using vehicle dynamics modelling tools has expanded to include many different design objectives, and has even influenced the vehicle development process. The principle of *model-based design,* in which system performance is predicted early enough in the design process to affect design choices, has seen application to the areas of vehicle handling, acceleration performance, ride quality, and durability of suspension components. The relatively recent addition of driver aids, such as anti-lock braking and stability control, are based on application of feedback control theory to vehicle dynamics models. Vehicle dynamics will only play an increasingly important role in the future, with expanded on-board computing for integrated active safety systems, with built-in mathematical estimation and feedback control techniques being widely deployed. Automated lanekeeping and emergency braking, and autonomous vehicles, are now in active development and on the verge of widespread deployment.

References

1 Olley, M., 1946. Road manners of the modern car. *Proceedings of the Institution of Automobile Engineers*, 41(1), pp. 523–551.
2 Segel, L., 1956. Theoretical prediction and experimental substantiation of the response of the automobile to steering control. *Proceedings of the Institution of Mechanical Engineers: Automobile Division*, 10(1), pp. 310–330.
3 Ellis, J.R., 1969. *Vehicle Dynamics*. Random House Business.
4 Sharp, R.S., 1971. The stability and control of motorcycles. *Journal of Mechanical Engineering Science*, 13(5), pp. 316–329.
5 Smith, C., 1978. *Tune to Win: The Art and Science of Race Car Development and Tuning*. Aero Publishers.
6 Wong, J.Y., 1978. *Theory of Ground Vehicles*. John Wiley & Sons.
7 Bakker, E., Nyborg, L. and Pacejka, H.B., 1987. *Tyre modelling for use in vehicle dynamics studies* (No. 870421). Society of Automotive Engineers Technical Paper.
8 Gillespie, T.D., 1992. *Fundamentals of Vehicle Dynamics*. Society of Automotive Engineers.
9 Ellis, J.R., 1994. *Vehicle Handling Dynamics*, Mechanical Engineering Publication Limited.
10 Milliken, W.F. and Milliken, D.L., 1995. *Race Car Vehicle Dynamics*. Society of Automotive Engineers.
11 Milliken, W.F., Milliken, D.L. and Olley, M., 2002. *Chassis Design: Principles and Analysis*. Society of Automotive Engineers.
12 Pacejka, H., 2005. *Tire and Vehicle Dynamics*. Elsevier.
13 Milliken, W.F., 2009. *Equations of Motion: Adventure, Risk and Innovation*. Bentley.

2

Tire Modelling

Many of the topics that will be discussed in this text are best explained by the development of linear dynamic models. However, there are a number of others that are not suitable to be resolved in this manner. Typically, this is due to nonlinearities in the equations that describe the resulting behaviour. These nonlinearities complicate matters, as they reduce the number of options available regarding mathematical solution techniques, and they often require the use of numerical methods to obtain solutions. Unfortunately, the tire is one of these elements. In fact, tires happen to be one of the most challenging vehicle components to characterize. While this complexity may be initially surprising, consider that the tire is a pressure vessel with relatively complicated geometry, experiencing external contact involving friction, large deflections, and centrifugal force effects. Add to this the nonlinear, anisotropic, and temperature dependant material properties, and the challenge becomes obvious. Nevertheless, despite this challenge, the tire is also one of the most important elements in any vehicle model, as it is the primary means of driver control over the vehicle motion.

The basic laws governing friction do not represent tire behaviour very well, but there are very many models for tire force prediction to choose from when reviewing the literature. In fact, there are entire texts devoted almost entirely to tire modelling – Pacejka [1] is an obvious example. As a consequence, this text is not intended to serve as a comprehensive source of tire modelling information, but rather will simply introduce the reader to the basics.

2.1 Rolling Losses

The first point of interest is *rolling resistance*. When a tire is rolling on the road, there are unavoidable energy losses that occur. Rolling resistance is often represented as a rolling resistance force that acts on the tire over the contact area with the road (commonly referred to as the contact patch). It is typically in

Fundamentals of Vehicle Dynamics and Modelling: A Textbook for Engineers with Illustrations and Examples, First Edition. Bruce P. Minaker.

Companion website: www.wiley.com/go/minaker/vehicle-dynamics

the order of 2% of the normal force that the tire supports. While this simple approximation is useful for a quick calculation, it is also a frequent source of confusion, when one tries to combine a rolling resistance force with a traction force: how can a tire simultaneously generate forces in both longitudinal directions? The fact that simple sliding friction always acts in a direction that opposes relative motion is a further complication, and often leads students to erroneous conclusions. In fact, tire traction is also friction force, and for a driving wheel, will act on the tire in the direction of vehicle motion.

A more accurate representation of the state of the tire contact forces is obtained by considering the rolling resistance not as a force, but rather as a moment that acts against the rotation of the wheel and tire. This moment is applied through the normal forces that act on the tire, and their distribution over the contact patch. As the tire material enters the contact patch, it compresses radially, and when it leaves the contact patch, it recovers this radial compression. However, due to hysteretic effects, the material does not behave identically while compressing and recovering, with more force required during the compression phase. Figure 2.1 illustrates a typical nonsymmetric normal force distribution that results when the tire is in motion. The centre of pressure of the normal force shifts forward, causing a moment in the direction opposite to rotation, resulting in a rolling resistance. Once the tire comes to rest, the pressure distribution returns to a symmetric shape, and the rolling resistance moment disappears. Rather than attempt to model this pressure force distribution, a simple rolling resistance force based on the normal force Z can be used. This rolling resistance force is effectively the force that must be applied to the wheel centre by the axle to balance the resistance moment, and as a result, an equal and opposite force acts on the vehicle to resist motion.

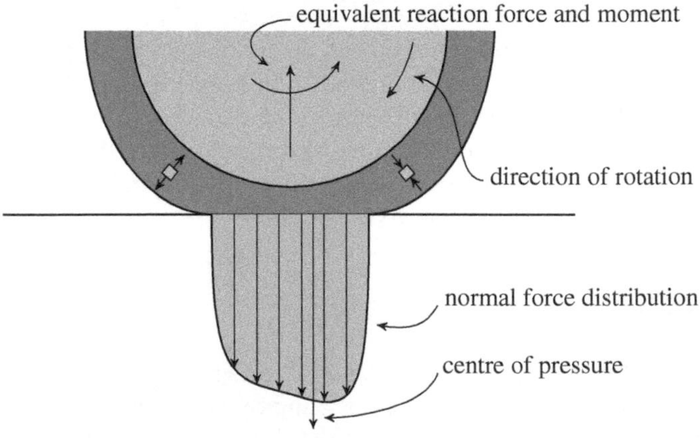

Figure 2.1 Tire rolling resistance is caused by a nonsymmetric normal force distribution between the tire and the road. The centre of pressure moves forward in response to rotation of the tire, causing a resistance to rotation.

The rolling resistance force of the tire X_t can be computed using a rolling loss coefficient k_r.

$$X_t = k_r Z \tag{2.1}$$

The rolling loss coefficient is a function of the forward speed of the tire u_t.

$$k_r = c_0 + c_1 u_t^2 \tag{2.2}$$

This expression is only an approximation of the true behaviour, is based on experimental results, and should be fitted with measured data in order to get results for a particular tire. However, Jazar [2] states that values of $c_0 = 0.015$ and $c_1 = 7 \times 10^{-6}\ \mathrm{s^2/m^2}$ are suitable in general, while Genta [3] gives values of $c_0 = 0.013$ and $c_1 = 6.5 \times 10^{-6}\ \mathrm{s^2/m^2}$, and Wong [4] gives values of $c_0 = 0.0136$ and $c_1 = 5.2 \times 10^{-7}\ \mathrm{s^2/m^2}$.

2.2 Longitudinal Force

While somewhat counterintuitive, it is now well understood that due to their flexibility, tires do not roll with an angular speed that is easily related to the vehicle forward speed. Figure 2.2 illustrates how the traction force generated by the tire causes it to compress tangentially as it enters the contact patch, and stretch as it leaves. As a result, the rotation of the tire is larger than would be the case if it were treated as a rigid body. In the case where a tire is acting under braking moments, the rotation is less than expected. A means of quantifying tire longitudinal slip is the *slip ratio* σ, which is defined as:

$$\sigma = \frac{\omega r_e - u_t}{u_t} = \frac{\omega r_e}{u_t} - 1 \tag{2.3}$$

where ω is the rotational speed of the tire, and r_e is the effective radius. An examination of the slip ratio shows that if the tire does roll as a rigid body,

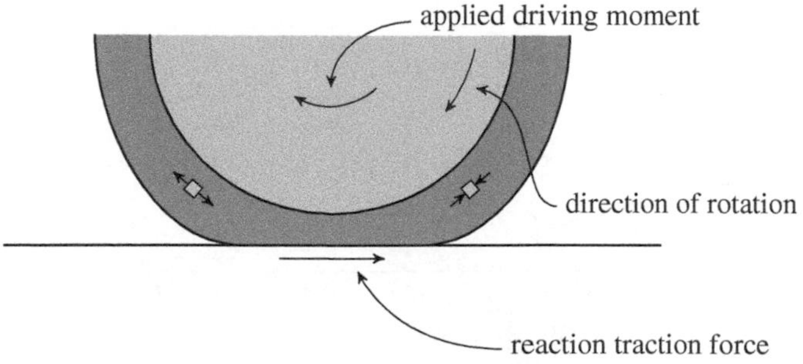

Figure 2.2 The slip ratio σ is defined as the ratio of the slip speed at the contact patch to the forward speed of the tire. The slip is caused by tangential compression of the tire as it enters the contact patch, and tangential stretch as is leaves.

i.e., where $u_t = \omega r_e$, then the slip $\sigma = 0$. If the wheel is locked and skidding as in a panic braking situation, then $\omega = 0$, and $\sigma = -1$. In order to have a $\sigma < -1$, the wheel must rotate against the direction of travel, a most unusual situation. In most normal driving situations, $0 < \sigma < 0.1$. If the wheel is spinning very quickly, as it does when it breaks loose under hard acceleration, then it is possible for $\sigma > 1$. The effective radius r_e does not necessarily correspond exactly to the physical radius, as this varies significantly due to normal load. The effective radius is defined more precisely as the value that satisfies $u_t = \omega r_e$ when the wheel is rolling freely, i.e., with no driving or braking moments, but the physical dimension is a good approximation.

Note that in the case of wheelspin for a stationary vehicle, the slip ratio goes to infinity. To prevent this, some authors will use an alternate definition of slip ratio, where the denominator is replaced by ωr_e in the case that $\omega r_e > u_t$. This change limits the maximum slip ratio to $\sigma = 1$, but has little effect otherwise. While the traction force generated by the tire is a function of many variables, it is very strongly dependent on the slip ratio. Figure 2.3 illustrates a typical longitudinal force versus slip ratio relationship. In the range of small slip ratios, the relationship is approximately linear, but clearly there is a maximum traction force that can be delivered, and at high slip ratios, the traction force depends strongly on the coefficient of friction μ of the tire against the road. For most street tires, the maximum grip will be in the vicinity of $\mu \approx 0.8$, but this is a function of tire and road wear condition, temperature, and weather.

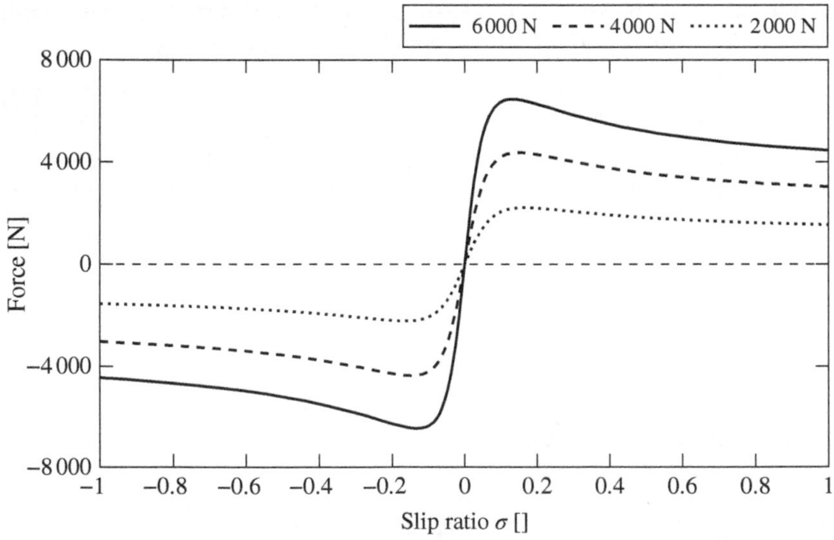

Figure 2.3 Tire longitudinal forces are primarily a function of slip ratio σ, but also depend on normal force, especially at high slip ratios. Here, tire longitudinal force is plotted against slip ratio for three different normal loads. The curves are not from a particular tire but are representative of a typical behaviour.

2.3 Lateral Force

In a situation similar to that experienced when the tire is loaded in the longitudinal direction, a slip phenomenon also governs the generation of lateral tire forces. The tire must deflect laterally in response to a lateral force. As the tire rolls, an unloaded, undeflected section of the tire passes into the contact patch, where it assumes a portion of the load. Its lateral deflection increases as it moves through the contact patch, until the normal force is insufficient to maintain the necessary level of grip with the road. As the deformed tire leaves the contact patch, it quickly resumes its undeformed shape. As a result, the shape of the lateral force distribution is not uniform or symmetric, but rather approximately triangular, with the rate of lateral deflection being related to the rate of lateral motion of the tire. The effect is illustrated in Figure 2.4. In effect, the tire does not move in the direction that it is pointed, but rather it 'creeps' sideways as it rolls forward. The angle between the direction that the tire is pointed, and the direction it travels is called the *slip angle* α. In normal operation, the slip angle is usually very small, less than 5°, and not immediately noticeable to the naked eye. The slip angle can be computed from:

$$\tan\alpha = \frac{v_t}{u_t} \tag{2.4}$$

where, as before, u_t is the forward speed of the tire, and v_t is the lateral speed.

As the slip angle increases, the slope of the triangular force distribution will increase, increasing the area under the triangle representing the total lateral force generated. The lateral force is therefore initially proportional to the slip

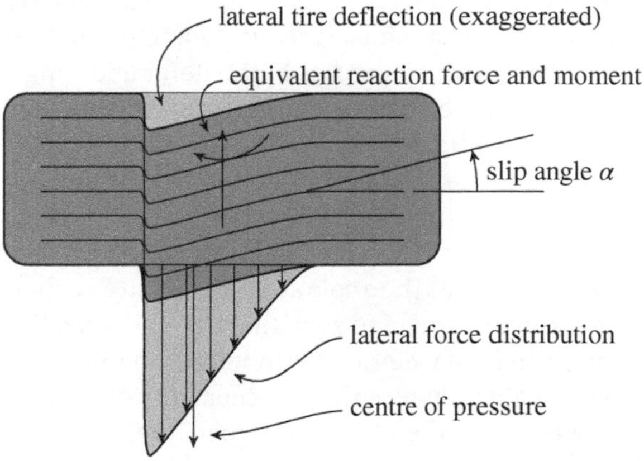

Figure 2.4 The slip angle α is defined as the angle between the direction the tire points, and the direction it travels. The ratio of the lateral speed to the forward speed of the tire will determine the slip angle. The lateral deflection of the tire increases approximately linearly from zero as it passes through the contact patch, with a rapid recovery as it leaves. This deflection causes the lateral motion of the tire, and the associated slip angle.

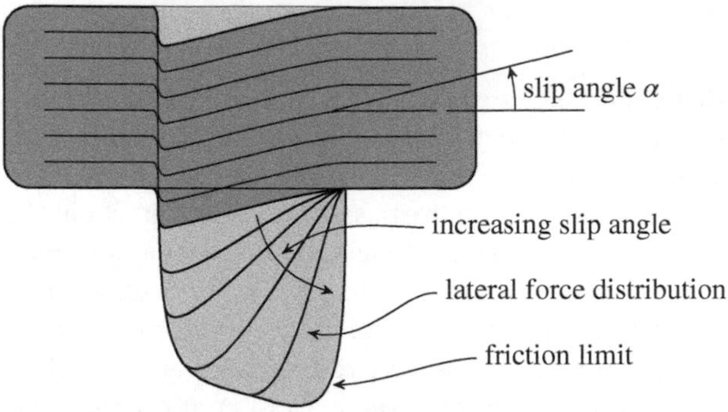

Figure 2.5 The triangular shape of the lateral force distribution holds only at small slip angles. As the total lateral force increases, the shape of the distribution is increasingly influenced by the normal force distribution, pushing the centre of pressure forward.

angle. Of course, the maximum lateral force is ultimately governed by the normal force and the friction with the road. As a result, the friction limit will vary along the contact patch with a distribution shaped similarly to the normal force distribution. Because the lateral forces will reach their limit of grip at the rear side of the contact patch first, and progressively fill the available space as the total lateral force is increased, the triangular shape of the lateral force distribution will change at high slip angles, and the total lateral force will no longer increase proportionally to the slip angle. The effect is illustrated in Figure 2.5.

Figure 2.6 illustrates a typical lateral force response curve as a function of slip angle. Typically, lateral and longitudinal forces behave somewhat similarly in response to slip. In fact, one of the most widely used tire force prediction models uses expressions of the same form to predict both lateral and longitudinal force. The slope of the lateral force curve with respect to slip angle is known as the *cornering stiffness*; this property has a significant effect on vehicle handling, and will be discussed in more detail in Chapter 4. Typically the response curve is symmetric, but can be influenced by other factors, such as *camber angle*, *conicity*, and *ply steer*.

The camber angle ε of the tire is defined as the angle at which the tire is tilted relative to the vertical. Negative camber is defined such that the top of the tire is inclined toward the vehicle, regardless of whether it is mounted on the left or right side of the vehicle. Camber effects influence the grip capability of the tire. Generally, the more contact area between the tire and the road, the better the grip. Because the tire will deform slightly in response to lateral loads, a small negative camber, in the order of $-2°$, tends to improve the grip capability of the

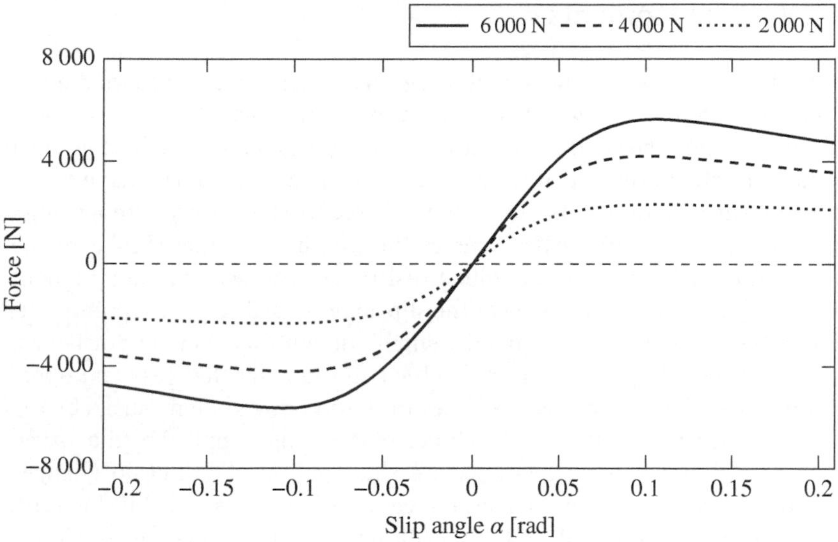

Figure 2.6 Tire lateral forces are primarily a function of slip angle α, but also depend on normal force. At low slip angles, they are approximately proportional to slip angle; the slope of the response curve is called the cornering stiffness.

outer tire, which is generally more highly loaded in both the vertical and lateral directions. If the camber angle is positive or largely negative, the contact area will reduce or become irregularly shaped, and grip will be reduced. Wider tires tend to be more sensitive to the effect. In single track vehicles like bicycles and motorcycles, where the profile of the tire is purposely rounded to accommodate large roll angles, there are more significant camber effects. Forces generated due to camber angle, called *camber thrust*, also contribute to the total lateral force. In these cases, the camber thrust sensitivity is relatively much smaller than the cornering stiffness, but the total lateral force will be due to both camber angle and slip angle effects.

Conicity and ply steer are the results of slight manufacturing defects in the tire construction. The term conicity refers to a slight conical shape of the tire, where ply steer refers to a small misalignment in the angles at which the layers, or plys, of the tire are arranged. Both effects can result in small nonzero lateral forces being present when the slip angle is zero, and it can be difficult to distinguish which of the effects is the cause. However, a simple reversal of the rotational direction of the tire is helpful here; the effects of conicity are independent of the direction of rotation, while the ply steer forces reverse direction with the rotation.

2.4 Vertical Moments

The mechanism by which the lateral forces are generated also causes the tire to generate *self-aligning moments*. The centre of pressure of the lateral force tends to be behind the centre of the tire, breaking the contact patch length in a two thirds to one third ratio, due to the approximately triangular shape of the lateral force distribution. This shift, often referred to as *pneumatic trail*, results in a moment around the vertical axis of the tire, in a manner similar to the way that normal forces generate rolling resistance. This self-aligning moment tends to act in a manner that reduces the slip angle, explaining the naming convention. These moments are generally small enough that they do not have a significant direct effect on the overall vehicle motion, but they play an important role in driver feedback, as the forces in the steering system caused by the aligning moments can influence the choice of steer angle applied by the driver.

As the total lateral force increase, the shape of the distribution changes away from triangular, and as a result, the centre of pressure will shift forward, reducing the pneumatic trail, and the resulting moment. Unlike the longitudinal and lateral forces that tend to constant values at large slip, the self-aligning moment tends back to zero or even small negative values. Very experienced drivers can use this feedback information to detect when the steering effort is no longer increasing with steer angle to recognize that the tire is approaching

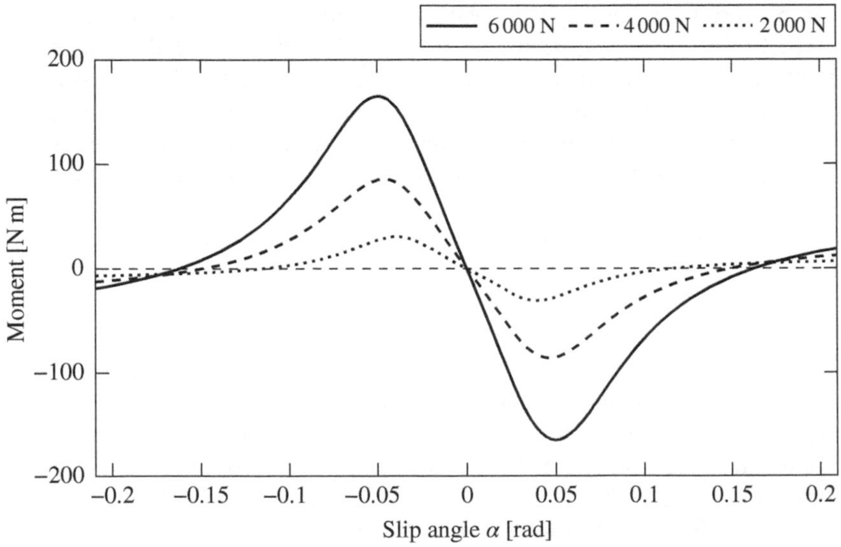

Figure 2.7 Tire self-aligning moments are primarily a function of slip angle α, but also depend on normal force. Unlike the lateral and longitudinal forces, they tend to zero at large slip angles.

the maximum level of grip. A typical aligning moment vs slip angle plot is shown in Figure 2.7.

2.5 Normal Force

The normal force between the tire and the road is a strong function of the vertical compression. The tire behaves much like a spring between the wheel and the ground. The vertical stiffness of a typical tire is about 150–200 kN/m but depends strongly on the inflation pressure; the mechanical stiffness of the tire only accounts for about 10% of the total vertical stiffness. The stiffness tends to increase slightly with compression, but the effect is weak. The damping of a tire in the vertical direction is relatively small, to the point where it is often ignored in mathematical models, with minimal influence on the result.

Figure 2.8 shows the effect of normal load on the grip capability of a tire. An increase in normal load will increase the grip, but with decreasing sensitivity. In effect, the coefficient of friction is reduced with increasing normal load. This leads to an important effect when a vehicle is cornering. The centripetal acceleration causes a rolling moment to be applied to the vehicle as it corners. To counter this roll moment, there is a shift in normal loads commonly referred to as *weight transfer*. The normal load on the inner tires is reduced to below the static level, while the outer tires carry more load. As the inner tires become

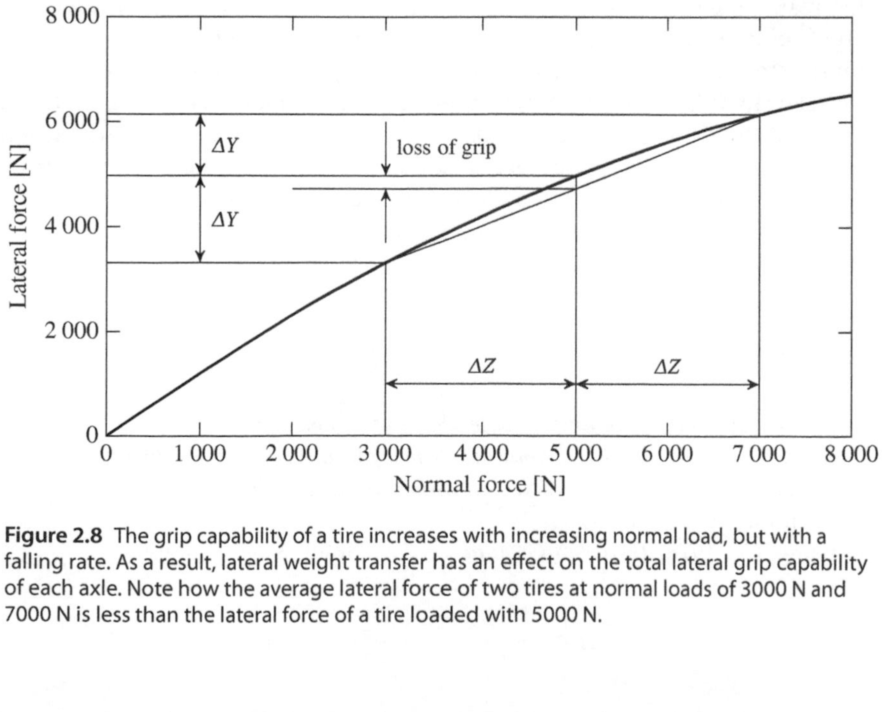

Figure 2.8 The grip capability of a tire increases with increasing normal load, but with a falling rate. As a result, lateral weight transfer has an effect on the total lateral grip capability of each axle. Note how the average lateral force of two tires at normal loads of 3000 N and 7000 N is less than the lateral force of a tire loaded with 5000 N.

unloaded, their ability to generate cornering forces is reduced; simultaneously, the grip capability of the outer tires is increased, but not enough to completely compensate. As a result, the grip capability of each axle as a whole is reduced by weight transfer.

The total amount of weight transfer experienced by the vehicle is determined by several factors: the lateral acceleration, the height of the mass centre of the vehicle, and the track width (the width of the vehicle, measured between the central planes of a left and right side tire). As a general rule, vehicles with lower centres of mass and wider track widths will be better able to generate cornering forces. Because the roll angle is approximately proportional to the lateral acceleration, as is the weight transfer, it is a common misconception that increasing the roll stiffness will reduce the lateral weight transfer, but this is not the case. Increased roll stiffness can reduce the roll angle, but not the weight transfer. (There is a component of weight transfer that occurs due to the lateral shift of the mass centre as the chassis rolls, which can be reduced by increasing roll stiffness, but this is usually only a small effect.)

A more important factor is the *relative* weight transfer, i.e., the ratio of weight transfer between the front and rear axles, which is is a function of the *relative roll stiffness* of the two axles. Consider the case of a vehicle with a very high roll stiffness in the front suspension, and comparatively low roll stiffness in the rear suspension. In this case, the majority of the weight transfer will occur across the front axle, decreasing its grip capability relative to the rear suspension. This can cause an effect known as *understeer*, where the front axle loses grip before the rear. Similarly, the opposite case, with a high rear roll stiffness, can cause an effect known as *oversteer* as the grip capability of the rear axle is reduced under lateral load. As a result of this effect, changes in roll stiffness can influence overall vehicle handling. In fact, in a racing environment, it is a common tactic for the pit crew to make exactly these type of adjustments to the suspension settings, in response to driver feedback that the vehicle is 'loose' (oversteering) or 'pushing' (understeering). There are more precise mathematical definitions of these two terms presented later in the text, when a detailed handling model is presented, but the general meaning is largely the same.

References

1 Pacejka, H., 2005. *Tire and Vehicle Dynamics*. Elsevier.
2 Jazar, R.N., 2008. *Vehicle Dynamics: Theory and Application*. Springer.
3 Genta, G., 1997. *Motor Vehicle Dynamics: Modeling and Simulation*. World Scientific.
4 Wong, J.Y., 1978. *Theory of Ground Vehicles*. John Wiley & Sons.

3

Longitudinal Dynamics

As the name implies, longitudinal vehicle dynamics deals with acceleration and braking performance. Although similar in nature, the two cases are typically treated separately. One of the primary means of assessing vehicle performance is a measurement of its acceleration performance. There are two regions in which the acceleration performance faces different limiting factors. At low speed, maximum acceleration is limited by tire grip. Most passenger vehicles produce a sufficient amount of torque in low gear to exceed the maximum grip capability of the tire. Thus, for most vehicles, at low speeds, the acceleration is independent of engine and drivetrain specifications. However, once even moderate speeds are reached, much less torque is available, and the performance is now limited by the engine and drivetrain. In the braking situation, this transition between regimes does not occur, and tire performance is generally the limiting factor for the duration of a braking event. Additionally, most passenger vehicles are driven by either the front or rear axle alone, while the brakes act on all four wheels.

3.1 Acceleration Performance

The longitudinal acceleration of a vehicle can be predicted reasonably accurately with simple single body model, using the equations of motion.

$$\sum \boldsymbol{f} = m\boldsymbol{a}_{\mathrm{G}} \tag{3.1}$$

In this case, only the longitudinal components of the force and acceleration vectors are needed.

$$\sum X\hat{\mathbf{i}} = m\dot{u}\hat{\mathbf{i}} \tag{3.2}$$

The primary longitudinal forces that act are the traction force developed by the rear axle, X_{r}, the traction force at the front axle, X_{f}, or both, the rolling resistance force X_{t}, the aerodynamic resistance force X_{a}, and the incline force, due to

Fundamentals of Vehicle Dynamics and Modelling: A Textbook for Engineers with Illustrations and Examples, First Edition. Bruce P. Minaker.

Companion website: www.wiley.com/go/minaker/vehicle-dynamics

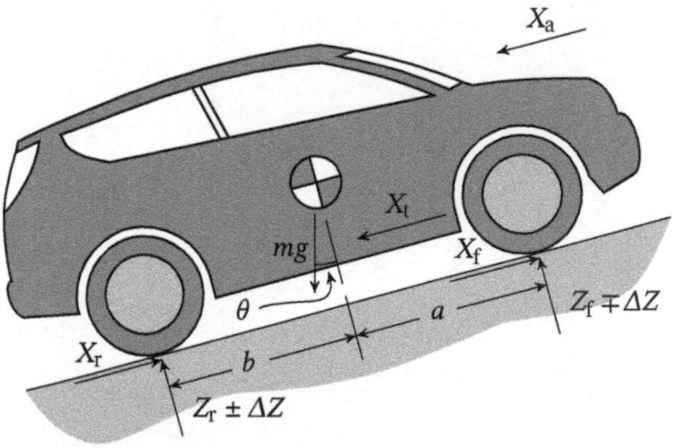

Figure 3.1 The longitudinal acceleration is based on a sum of traction and resistive forces.

any component of the weight that may be acting in the longitudinal direction. A free body diagram showing the relevant forces is shown in Figure 3.1. One may include load forces due to trailers where appropriate, if the analysis concerns trucks. Dropping the vector notation and substituting the various forces gives:

$$X_r + X_f - X_a - X_t - mg\sin\theta = m\dot{u} \tag{3.3}$$

One of the challenges when solving the equation of motion in this case is the nonlinearities that are present. The traction forces are a function of the state of the tire slip, which is influenced by the engine torque, which is in turn a complicated function of engine speed. The aerodynamic and tire rolling resistance forces depend on the speed, but in a nonlinear fashion, growing with increasing speed. Because of the way in which the forces vary, the most accurate approach for solving the longitudinal acceleration problem is a numerical solution. The equations can be cast as a first order differential equation, and solved using a numerical method.

3.1.1 Engine Limited Performance

When determining the engine limited performance of a road vehicle, there are two simple relations describing the drivetrain that will usually provide a sufficiently accurate result. The first relevant model is a simple kinematic relationship between the vehicle speed u and the engine rotational speed ω_e.

$$u = \frac{\omega_e r_e}{n_g n_{fd}(1+\sigma)} \tag{3.4}$$

The gear ratio of the gearbox is given as n_g; similarly, the gear ratio in the final drive (usually on the ring gear of the differential) is given as n_{fd}. The value is given as the ratio of input speed to output speed, or equivalently as the ratio of output torque to input torque. The gearbox reduction will typically be in the range of 3:1–5:1 in the lowest gear, and approximately 1:1 in top gear. The final drive will typically be in the 3:1–4:1 range. While traditionally the final drive ratio was fixed, in some of the newer double clutch gearboxes, it may also change depending on the gear selected. The engine speed is typically given in units of revolutions per minute (rpm), and usually operates in the range of 1000–6000 rpm, but must be converted to rad/s for use in Equation (3.4). The tire slip ratio σ may be treated as a constant in the range of 0.01–0.02 without significant error in most cases. The effective rolling radius of tire is denoted r_e; a value close to 0.3 m would be normal for a passenger car.

To accompany the driveline kinematics, there is an associated relationship between engine torque t_e and traction force. Assuming the rear wheel drive (RWD) case, where the traction force acts at the rear wheel:

$$X_r = \frac{t_e n_g n_{fd} \eta}{r_e} \tag{3.5}$$

The torque produced by the engine is typically in the range of 100–400 Nm. The efficiency of the drivetrain is given as η, and for most manual shift gearboxes, is approximately 0.9. A fixed efficiency is only an approximation, as many of the losses are not proportional to the torque transmitted. Using these relationships, and given an engine performance curve, the traction force available can be plotted as a function of vehicle speed, as shown in Figure 3.2. Using this relation to compute the traction force available to accelerate the vehicle is slightly

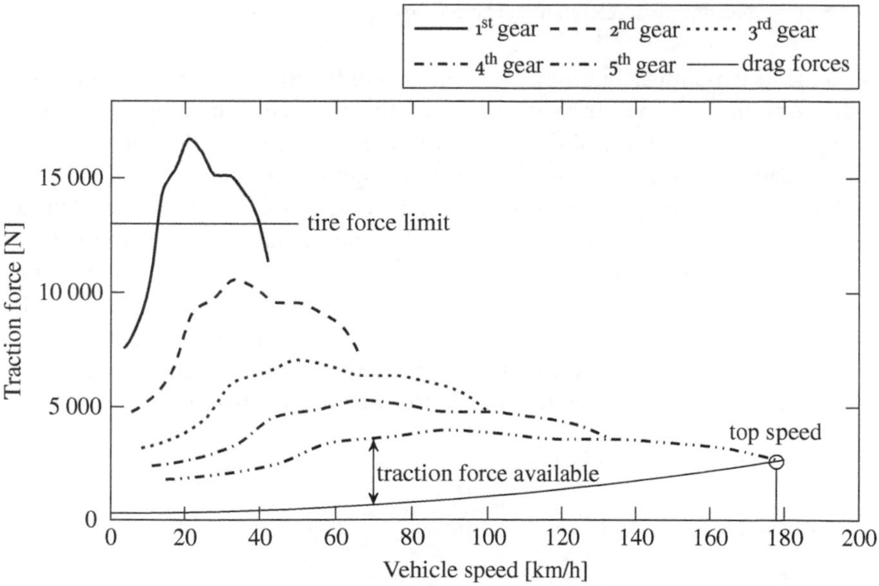

Figure 3.2 Traction force available from the tire can be plotted as a function of vehicle speed. Note the difference between the traction force and drag forces is visible in the plot, and is a measure of the available acceleration performance. The drag force and traction force intersect at the vehicle's top speed.

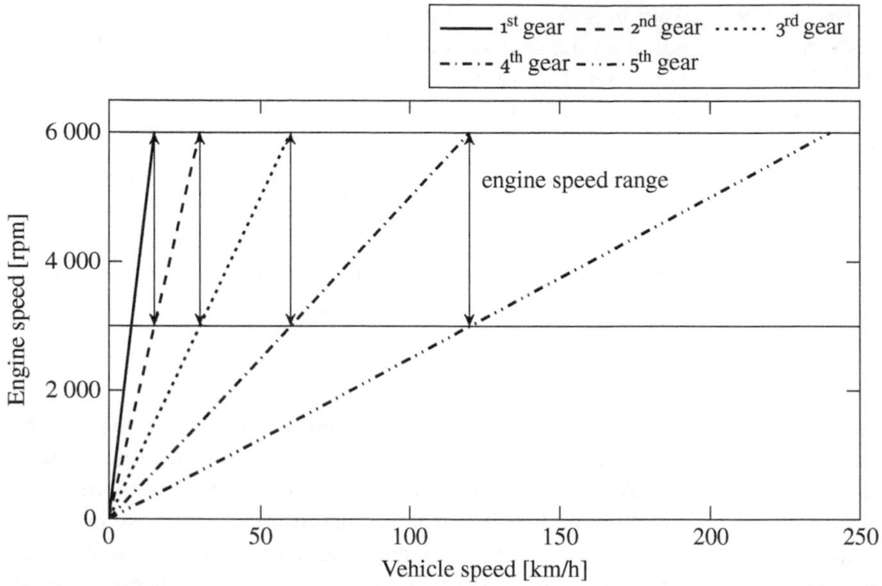

Figure 3.3 The gear ratios can be selected so the engine always operates in a fixed range of speeds, but this requires an increasingly wider speed range for each gear. This strategy is not optimal for maximizing acceleration performance in passenger cars.

in error, as it assumes that the angular inertia of the drivetrain can be neglected relative to the inertia of the vehicle. While this is usually an acceptable simplification, the relative size of the drivetrain inertia is strongly dependent on the drive ratio, and in low gear, it is much more significant.

A challenge that is often faced by vehicle design engineers is the selection of the drivetrain ratios to maximize vehicle performance. There are various strategies employed, e.g., one can choose the ratios such that the engine speed is maintained within a specified operating range, as shown in Figure 3.3. The ratios follow a *geometric progression*. An inspection of the figure shows that the range of vehicle speeds increases quite dramatically as the ratios progress. In this case, the ratio of successive gear ratios must be a constant.

$$\frac{n_2}{n_1} = \frac{n_3}{n_2} = \frac{n_4}{n_3} = \cdots = k_g \tag{3.6}$$

This implies:

$$n_{g+1} = k_g n_g \tag{3.7}$$

or:

$$k_g = \left(\frac{n_{g+1}}{n_1}\right)^{1/g} \tag{3.8}$$

As long as any two ratios are determined, the remaining ratios can be computed. This strategy is best employed on heavy vehicles such as highway tractor trailers, where there are often very many ratios utilized (15–20). For a passenger car, this strategy does not give optimal acceleration performance. A significant increase in higher gear performance can be obtained with a small penalty in decreased low gear performance, by increasing the spread in the lower ratios. A strategy along these lines is to set the range of speeds covered in each gear as a constant. In this case, there is a relationship between maximum speeds in each gear.

$$u_{\max_{g+1}} - u_{\max_g} = u_{\max_g} - u_{\max_{g-1}} \tag{3.9}$$

Substitution of Equation (3.4) in Equation (3.9) gives a relationship between successive ratios.

$$n_{g+1} = \frac{n_g n_{g-1}}{2n_{g-1} - n_g} \tag{3.10}$$

This strategy is illustrated in Figure 3.4.

One important point to note is that the vehicle top speed is only indirectly influenced by the choice of gear ratio. The maximum achievable vehicle speed is ultimately dependent on the engine power. However, a poor selection of the top gear ratio can prevent the vehicle from reaching this maximum achievable

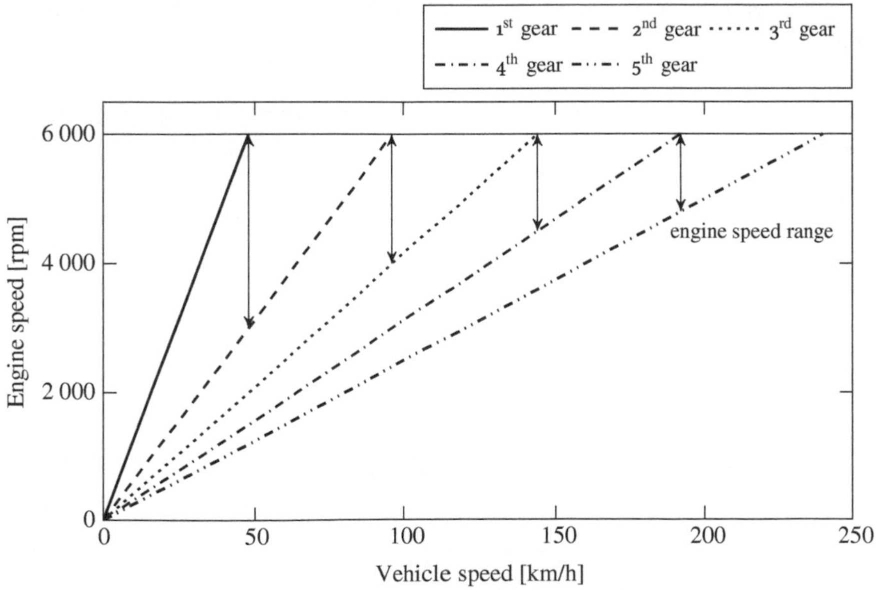

Figure 3.4 The gear ratios can be selected so the range of speeds remains constant. The range of engine operating speeds decreases as the gears progress.

top speed. If the top gear ratio is too low (numerically large, giving more angular speed reduction), the vehicle will reach the maximum safe engine speed before reaching the highest achievable vehicle speed. The vehicle is said to be *undergeared*. Conversely, if the top gear is too high, then the traction force produced will be reduced, lowering the top speed, and preventing the engine from reaching the speed range where power is maximized. In this case, the vehicle is said to be *overgeared*. An optimal choice of top gear ratio will allow the vehicle to reach its maximum top speed. In a passenger car, the driver is typically more concerned with fuel economy than top speed, and so the top gear is often chosen such that the engine operates in its most efficient range of speeds when the vehicle is at highway speed. This choice typically results in the vehicle being overgeared, and explains why the measured top speed of many passenger cars is higher in the gear below top gear.

Example

Consider a vehicle with a final drive ratio of 3.15:1, and a wheel with an effective rolling radius of 0.3 m. The engine fitted to the vehicle makes a maximum power of 150 kW at 5000 rpm. The aerodynamic resistance can be modelled as:

$$X_a = \frac{1}{2}\rho A_f c_d u^2 \tag{3.11}$$

where ρ, A_f, and c_d are the air density, the frontal area, and the coefficient of drag, respectively. The rolling resistance loads can be modelled as shown in Figure 3.5.

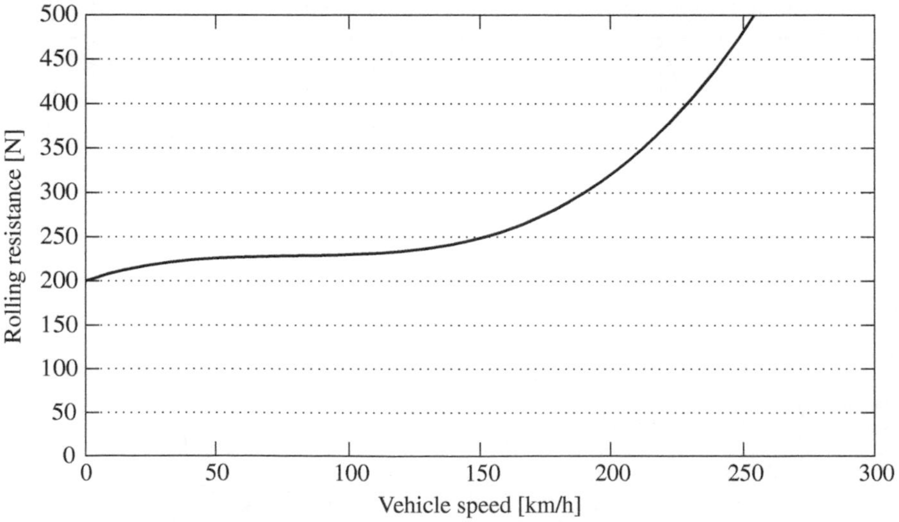

Figure 3.5 Rolling resistance forces vs vehicle speed.

If the frontal area $A_f = 2.2\ m^2$ and the drag coefficient is $c_D = 0.3$, compute the optimal top gear ratio for maximum top speed. Assume that the density of air is $\rho = 1.23\ kg/m^3$, and that tire slip is about 5%. The power required to overcome resistance forces is:

$$P = (X_a + X_t)u \tag{3.12}$$

Estimating a top speed of 200 km/h = 55.6 m/s gives total resistance losses of:

$$X_a + X_t = \frac{1}{2}(1.23)(2.2)(0.3)(55.6)^2 + 325 = 1578\ \text{N}$$

Calculating the power to overcome the aerodynamic and rolling resistance gives:

$$P = (X_a + X_t)u = (1578)(55.6) = 87654\ \text{W}$$

Comparing the required power to the power available (150 kW) shows that the top speed is well above 200 km/h. Recomputing at 250 km/h gives a power requirement of 169 kW. Clearly the top speed falls between these two values. Iteration leads to a convergence at approximately 241 km/h. Assuming the vehicle hits its top speed at precisely the engine speed where maximum power is delivered gives:

$$u = \frac{\omega_e r_e}{n_g n_{fd}(1+\sigma)} = \frac{241}{3.6} = \frac{5000\left(\frac{2\pi}{60}\right)0.3}{(n)(3.15)(1+0.05)}$$

Solving gives a value of $n = 0.709{:}1$. Typically, this would be modified to a slightly lower (but numerically higher) ratio, in the order of 5%, i.e., $n = 0.74$, to put the engine speed slightly above the point where maximum power is generated (i.e., slightly undergeared). In this case, when the vehicle is running at maximum speed, a slight increase in load from wind or road grade will cause the engine to slow slightly, and increase its power output, offsetting the additional load. In the case of overgearing, a slight increase in load will slow the engine, and cause a reduction in power output, which will further reduce the top speed. As a result, in vehicles that are overgeared, the top speed may vary significantly based on the test conditions. An undergeared vehicle will be less susceptible to these effects.

3.1.2 Tire Limited Acceleration

At low speed, the maximum acceleration will be limited by the tire grip available. Consider the sum of forces acting on the vehicle.

$$\sum f = m a_G \tag{3.13}$$

In the horizontal direction, the traction forces act to generate longitudinal acceleration. In this condition, the aerodynamic drag and rolling loss loads can be ignored, as the speeds at which acceleration is governed by tire behaviour are generally slow enough that they are negligible.

$$\sum X = X_\mathrm{f} + X_\mathrm{r} = m\dot{u} \tag{3.14}$$

In the vertical direction, the vehicle is assumed to be in equilibrium. The normal forces are written as a sum of the static loads, plus the shift that results when the vehicle accelerates.

$$\sum Z = Z_\mathrm{f} - \Delta Z + Z_\mathrm{r} + \Delta Z - mg = m\dot{w} = 0 \tag{3.15}$$

Consider the sum of moments acting on the vehicle.

$$\sum \boldsymbol{m}_\mathrm{G} = \mathbf{I}_\mathrm{G}\boldsymbol{\alpha} + \boldsymbol{\omega} \times \mathbf{I}_\mathrm{G}\boldsymbol{\omega} \tag{3.16}$$

Only the pitch moment is important in this case, and the vehicle is assumed to be in pitch equilibrium. This is a small source of error, as the vehicle will experience a short transient period where the pitch angle assumes a new equilibrium value under acceleration, but the simplification is usually acceptable.

$$\sum M = h_\mathrm{G}(X_\mathrm{f} + X_\mathrm{r}) + a(Z_\mathrm{f} - \Delta Z) - b(Z_\mathrm{r} + \Delta Z) = I_{yy}\dot{q} = 0 \tag{3.17}$$

Setting $\Delta Z = \dot{u} = 0$ allows the solution of the equilibrium normal forces.

$$Z_\mathrm{f} = mg\frac{b}{a+b} \tag{3.18}$$

$$Z_\mathrm{r} = mg\frac{a}{a+b} \tag{3.19}$$

Examination of the expressions for normal forces shows that a simple fraction of the weight acts on each axle. For example, shifting the centre of mass forward by increasing the distance b and decreasing the distance a by the corresponding amount increases the fraction of weight acting on the front axle. Substitution of the expressions for the equilibrium normal forces allows solution of the change in normal force due to acceleration.

$$\Delta Z = m\dot{u}\frac{h}{a+b} = (X_\mathrm{f} + X_\mathrm{r})\frac{h_\mathrm{G}}{a+b} \tag{3.20}$$

A simple slip-or-stick tire model ($X \leq \mu Z$) is used to simplify the resulting expressions. If the tire model is applied on the front wheel, i.e., front wheel drive (FWD), accelerating to the point of front wheelspin, and the rear traction force is set to zero, the limit of acceleration can be found.

$$X_\mathrm{f} \leq \mu(Z_\mathrm{f} - \Delta Z), \; X_\mathrm{r} = 0 \tag{3.21}$$

$$X_\mathrm{f} \leq \mu\left(mg\frac{b}{a+b} - X_\mathrm{f}\frac{h_\mathrm{G}}{a+b}\right) \tag{3.22}$$

After some simplification, an expression for the maximum traction force can be found.

$$X_f \leq \frac{\mu mgb}{a+b+\mu h_G} \tag{3.23}$$

Substituting the maximum traction force gives the resulting maximum tire limited acceleration.

$$\dot{u}_{FWD} \leq \frac{\mu gb}{a+b+\mu h_G} \tag{3.24}$$

Similarly, the traction limit can be fixed on the rear axle, and the front traction force ignored to find the limit for rear wheel drive (RWD).

$$X_r \leq \mu(Z_r + \Delta Z),\ X_f = 0 \tag{3.25}$$

$$X_r \leq \frac{\mu mga}{a+b-\mu h_G} \tag{3.26}$$

$$\dot{u}_{RWD} \leq \frac{\mu ga}{a+b-\mu h_G} \tag{3.27}$$

The expressions for the FWD and RWD case are similar, with one interesting feature. The terms in the denominator show that as the height of the centre of mass is increased, the maximum acceleration for a RWD vehicle increases, while for a FWD vehicle, the opposite occurs.

3.1.3 Grade Performance

Another measure of vehicle performance is the ability to climb hills, or *gradeability*. Both the engine and the tire will impose limits on the maximum grade that a vehicle can negotiate. Consider the tire limited acceleration, i.e., Equations (3.24) and (3.27), recast by dividing by the gravitational constant, to find the maximum acceleration in units of gravities, or 'g'.

$$\frac{\dot{u}_{RWD}}{g} \leq \frac{\mu a}{a+b-\mu h_G} \tag{3.28}$$

A free body analysis will show that the same expressions apply to the maximum achievable grade before tire grip is lost, e.g.:

$$\tan\theta_{RWD} \leq \frac{\mu a}{a+b-\mu h_G} \tag{3.29}$$

Road grade is often reported in terms of rise over run as a percentage, i.e., if the inclination angle is θ, then $100\tan\theta$ is the percent grade. A 100% grade is therefore equivalent to $\theta = 45°$; a grade of 25% would be exceptionally steep, and most road grades fall well below 10%.

This result can be compared with the engine limited gradeability. In this case, it is common to ignore resistance forces, as most steep grades are attempted with speeds low enough that the weight component is dominant, and to assume a steady climbing speed. Equating the vehicle weight component acting along the grade to the engine limited traction force allows calculation of the engine limited grade.

$$mg\sin\theta = \frac{t_e n_g n_{fd} \eta}{r_e} \tag{3.30}$$

Most modern passenger cars will have sufficient engine torque that when running in low gear, it is the tire rather than the engine that will limit gradeability.

3.2 Braking Performance

In the case of braking, many of the equations are the same as the case of tire limited acceleration, except the forces X_f and X_r will be considered positive when acting in the rearward direction. It would be possible to consider the forces as negative, but this complicates the slip-or-stick friction model, which assumes positive forces. The acceleration $\dot{u}$ is assumed to be negative. In this case, the resistance losses are ignored, as they are generally small when compared to braking forces. Unlike the case for acceleration, the braking forces do not vary significantly as a function of speed. The sum of longitudinal forces becomes:

$$\sum X = -X_f - X_r = m\dot{u} \tag{3.31}$$

If the weight transfer on to the front axle is taken as a positive value, the sum of vertical forces gives:

$$\sum Z = Z_f + \Delta Z + Z_r - \Delta Z - mg = m\dot{w} = 0 \tag{3.32}$$

The sum of moments equation also has to be slightly modified to account for the change in direction of positive traction forces.

$$\sum M = -h_G(X_f + X_r) + a(Z_f + \Delta Z) - b(Z_r - \Delta Z) = I_{yy}\dot{q} = 0 \tag{3.33}$$

The weight transfer equation can now be written as:

$$\Delta Z = -m\dot{u}\frac{h}{a+b} = (X_f + X_r)\frac{h_G}{a+b} \tag{3.34}$$

If one defines the μ as the minimum coefficient of friction required to support the braking forces requested, then expressions for the front and rear friction can be found. For the front:

$$\mu_f = \frac{X_f}{Z_f + \Delta Z} \tag{3.35}$$

and for the rear:

$$\mu_r = \frac{X_r}{Z_r - \Delta Z} \tag{3.36}$$

In order to maintain a constant braking 'efficiency', i.e., a constant fraction of the amount of friction used on the front axle, and substituting the expression for weight transfer, a linear relationship between the two braking forces can be found to be:

$$X_f\left(1 - \frac{\mu_f h_G}{a+b}\right) = \mu_f\left(mg\frac{b}{a+b} + X_r\frac{h_G}{a+b}\right) \tag{3.37}$$

or:

$$X_f(a + b - \mu_f h_G) = X_r\mu_f h_G + \mu_f mgb \tag{3.38}$$

Similarly, if one assumes a constant ratio of the available braking friction on the rear axle, the following relationship can be found:

$$X_r(a + b + \mu_r h_G) = -X_f\mu_r h_G + \mu_r mga \tag{3.39}$$

The two linear equations can be cast in a matrix that allows for solution of X_f and X_r.

$$\begin{bmatrix} a+b-\mu_f h & -\mu_f h_G \\ \mu_r h & a+b+\mu_r h_G \end{bmatrix}\begin{Bmatrix} X_f \\ X_r \end{Bmatrix} = \begin{Bmatrix} \mu_f mgb \\ \mu_r mga \end{Bmatrix} \tag{3.40}$$

Solving gives:

$$\begin{Bmatrix} X_f \\ X_r \end{Bmatrix} = \begin{bmatrix} a+b-\mu_f h & -\mu_f h_G \\ \mu_r h & a+b+\mu_r h_G \end{bmatrix}^{-1}\begin{Bmatrix} \mu_f mgb \\ \mu_r mga \end{Bmatrix}$$

$$= \frac{\begin{bmatrix} a+b+\mu_r h & \mu_f h_G \\ -\mu_r h & a+b-\mu_f h_G \end{bmatrix}\begin{Bmatrix} \mu_f mgb \\ \mu_r mga \end{Bmatrix}}{(a+b-\mu_f h)(a+b+\mu_r h) + \mu_f\mu_r h_G^2} \tag{3.41}$$

$$X_f = \frac{mg\mu_f(b + h_G\mu_r)}{(a+b) + (\mu_r - \mu_f)h_G} \tag{3.42}$$

$$X_r = \frac{mg\mu_r(a - h_G\mu_f)}{(a+b) + (\mu_r - \mu_f)h_G} \tag{3.43}$$

The two lines are shown plotted in Figure 3.6, with the intersection at the coordinates solved in Equations (3.42) and (3.43).

If one sets the brake force distribution so that both axles utilize the same fraction of the available friction, then they should simultaneously reach the limits of traction, i.e., the brakes on both axles will lock at the same level of brake

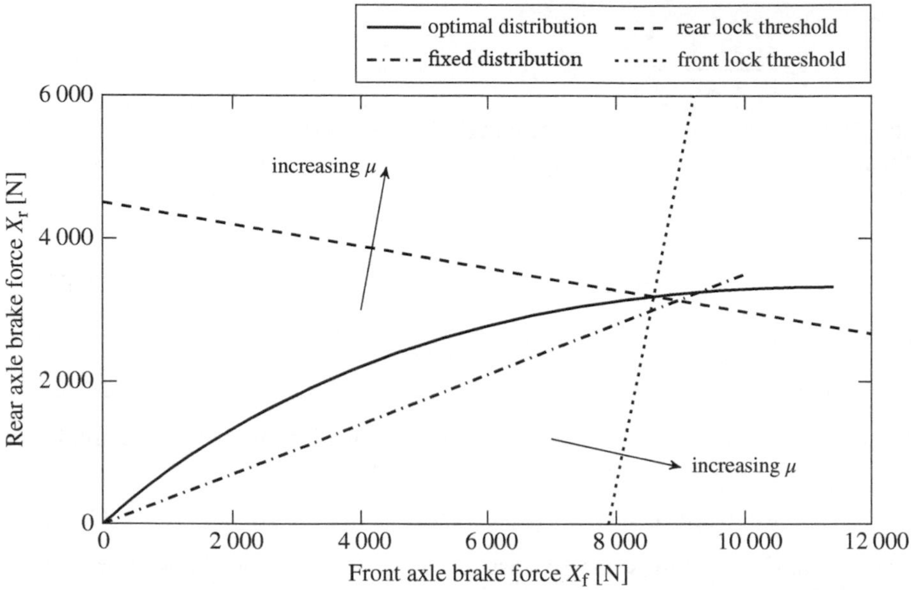

Figure 3.6 The optimal brake force distribution is a function of deceleration. The weight transfer that occurs during braking implies that a larger portion of the brake force should act on the front axle at higher values of deceleration in order to maintain a balanced friction load on both axles. In this example, the fixed distribution intersects the front threshold slightly before the rear threshold, indicating that the front tires would lock first. Unfortunately, if the friction were to increase, the threshold lines shift in a way that allows the rear lockup to occur first, a potentially dangerous situation. Plot generated using parameters in Table 3.1.

Table 3.1 Braking force distribution parameter values

Quantity	Notation	Value
front axle dimension	a	1.4 m
rear axle dimension	b	1.7 m
centre of mass height	h	0.7 m
mass	m	1500 kg
friction coefficient	μ	0.8 –

pedal pressure. The resulting forces at each axle are found as the locus of the intersection of the two linear relationships, as the level of friction is varied.

$$\mu_f = \frac{X_f}{Z_f + \Delta Z} = \mu_r = \frac{X_r}{Z_r - \Delta Z} \tag{3.44}$$

$$X_f(Z_r - \Delta Z) - X_r(Z_f + \Delta Z) = 0 \tag{3.45}$$

$$X_f\left(mg\frac{a}{a+b} - (X_f + X_r)\frac{h_G}{a+b}\right) - X_r\left(mg\frac{b}{a+b} + (X_f + X_r)\frac{h_G}{a+b}\right) = 0 \tag{3.46}$$

$$(X_f + X_r)^2 + \frac{mg}{h_G}(bX_r - aX_f) = 0 \tag{3.47}$$

The locus of the intersection indicates that the ideal distribution is given by a parabolic relationship between the braking force at the front and rear axles, as shown in Figure 3.6. This implies that the distribution should not remain fixed, but rather, an increasing fraction should be apportioned to the front axle as the braking force increases. This poses a problem, as typically the distribution is fixed and depends on the geometric properties of the hydraulic braking system, e.g., the brake cylinder and master cylinder diameters, the brake rotor diameter, the brake pad friction, etc.

A fixed relationship will rarely result in an optimal braking situation. If a vehicle is stopping with maximum deceleration, and the available friction is higher than anticipated during the design of the braking system, or if the vehicle is on a downward slope, there will be more weight transfer than expected, and the rear brakes will lock before the front. This situation is potentially very dangerous, as it eliminates the ability of the rear tires to generate any lateral forces, effectively reducing the rear cornering stiffness to zero. A study of the yaw plane model, covered in Chapter 4, will reveal that this condition results in a yaw instability. Even if the rear brakes are not entirely locked, at high levels of longitudinal force, the lateral force available is very low, and the result is effectively the same.

If the friction coefficient is very low, e.g., on a snow-covered or icy road, the level of deceleration will be very low, as will the weight transfer, and the front brakes will now lock first. While this does not result in an instability like the opposite case, it does eliminate the ability of the driver to steer the vehicle. In the case of a rear wheel drive vehicle fitted with an automatic transmission, in very slippery conditions, the front brakes may lock at a lower pedal force than is required for the rear brake force to overcome the drivetrain torques. In this situation, the driver is faced with a vehicle that does not respond to steer inputs, and yet is still being driven forward, even with the brake applied. (Quite often slippery road conditions coincide with cold weather, which causes the engine management system or carburetor to raise the idle speed as a result, further contributing to the problem.)

As can be seen, the design of the brake system must be a compromise; any changes to reduce one of the problems described above will make the other worse. In order to overcome this challenge, most road vehicles are now fitted with a brake pressure relief valve that limits or modifies the pressure in the rear hydraulic circuit, or alternatively, some sort of anti-lock or even a mechatronic braking system. An anti-lock braking system monitors the wheel speeds, and reduces the hydraulic pressure if too large an angular deceleration is detected, preventing the brake from locking. Mechatronic braking systems can use acceleration information as feedback to continuously readjust the brake force distribution.

Problems

3.1 Consider a truck with the following specifications:
Mass: 2300 kg
Wheelbase: 2.5 m
Centre of mass location: 1.0 m behind the front axle
Centre of mass height: 0.7 m above the ground
Drive axle ratio: 3.5:1
Tire rolling radius: 0.35 m

a) If the redline (maximum permissible engine speed) is 4500 rpm, and the maximum vehicle speed in first gear is 40 km/h, what must the first gear ratio be?
b) If the truck is rear wheel drive, what is the maximum acceleration it can achieve on a flat surface with coefficient of friction of $\mu = 0.8$, before wheelspin occurs? Assume the truck is moving slowly enough to ignore wind and rolling resistance. How much torque must the engine produce in order to spin the wheels in first gear on this surface?
c) If the engine can produce a maximum of 380 Nm of torque, what is the maximum grade it can climb? Is it limited by the engine torque, or the tire grip? Assume $\mu = 0.8$.
d) If the truck is travelling at 85 km/h, and the engine is running at 3000 rpm, while in fourth gear, what must the fourth gear ratio be? Assume that tire slip is 2%.
e) If the working speed range of the engine is the same in all gears, estimate the ratios of second and third gear.

3.2 Given a vehicle with the following specifications:
Four wheel drive, front engine, high output 5.9-liter turbo Diesel, six-cylinder inline
Weight: 3000 kg
Wheelbase: 3.5 m
Weight distribution (%f/r): 53/47
Centre of mass height: 1.0 m above the ground
Max power: 305 bhp (227 kW) @ 2900 rpm

Max torque: 555 lb-ft (752 Nm) @ 1400 rpm
Redline: 3200 rpm
Transmission: 6-sp. manual
Ratios: 5.63:1, 3.38:1, 2.04:1, 1.39:1, 1.00:1, 0.73:1
Axle Ratio 4.10:1
Tires: 265/70R-17
Assume that the truck is equipped with a central differential in its drivetrain, i.e., the front and rear axles must transmit the same amount of torque, and that the tire's coefficient of sliding friction is $\mu = 0.8$.

a) What is the maximum traction limited acceleration of the truck, if one assumes that front tire slip is the limiting factor? Ignore wind and rolling load in your calculation, giving some explanation of why this is (or isn't) reasonable. Does the front tire actually slip first? Hint: start with a free body diagram.
b) What is the maximum acceleration of the truck in first, second, and third gears? Can it reach the traction limit?
c) Judging by the choice of ratios that the design engineers have chosen for the transmission, comment on what you think was their objective in this choice.

3.3 Consider the snow plow shown in Problem 3.3. Assume that the snow loads on the plow can be treated as normal forces with a uniform distribution over the height of the plow, and that the blade is close to but does not contact the ground. The properties of the truck are as follows:
Mass: 2500 kg
Wheelbase: 3.2 m
Weight distribution (%f/r): 55/45
Distance from plow front edge to front axle: 1.5 m
Plow inclination: 30° from vertical
Plow height (measured parallel to plow surface) 1.0 m

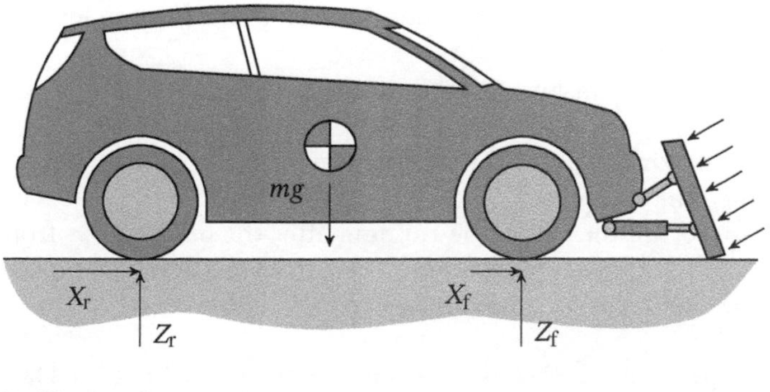

Problem 3.3 A snow plow.

a) Find the maximum amount of snow, in terms of traction force at the rear wheel, that can be pushed before the rear tire spins (assume rear wheel drive), with a coefficient of friction between the tire and the road of $\mu = 0.3$.
b) Repeat the above calculation, assuming that the truck is four wheel drive, with a drive system that ensures the traction force is the same at both axles.

3.4 Consider a vehicle with the following specifications:
Front engine, inline 6 cylinder
Length: 4488 mm (176.7 in)
Width: 1757 mm (69.2 in)
Height: 1369 mm (53.9 in)
Weight: 1505 kg (3318 lb)
Wheelbase: 2725 mm (107.3 in)
Weight distribution: (%f/r): 50/50
Centre of mass height: 0.5 m above the ground.
Max power: 225 bhp (168 kW), 5900 rpm
Max torque: 214 lb-ft (291 Nm), 3500 rpm
Gear ratios: 4.21:1, 2.49:1, 1.66:1, 1.24:1, 1.00:1
Final Drive: 2.93:1
Redline: 6500 rpm
Tires: 205/50R-17
Drag coefficient: 0.31
Frontal area: 2.2 m^2
Assume a coefficient of friction of $\mu = 0.8$, and that the rolling resistance loads can be modelled as shown in Figure 3.5.

a) Find the vehicle speed in third, fourth, and fifth gears, when the engine is at redline. Make appropriate assumptions for the efficiency and tire slip. Do you expect the car to reach redline in each of these three gears? If not, what speeds do you expect? You should give evidence for up your expected values with a calculation, but some small approximation error is acceptable.
b) What is the maximum low speed acceleration? What is the limiting factor?
c) What is the maximum grade that the car can negotiate in second gear? What is the limiting factor?
d) At what speed will the engine be running after the driver shifts from first to second at redline? What about from from third to fourth, (also at redline)? Comment on the significance of the difference.

3.5 A student competition vehicle has experienced a hydraulic brake line failure, disabling the front brakes of the vehicle. Luckily, the vehicle has been

designed with separate hydraulic circuits, and the rear brakes are still functional. The coefficient of friction of the surface is $\mu = 0.7$, and the vehicle properties are as follows:
Mass: 335 kg
Wheelbase: 1.7 m
Distance from the centre of mass to the front axle: 0.7 m
Height of the centre of mass: 0.6 m (above the ground)

a) What is the maximum deceleration that can be obtained?
b) What is maximum deceleration that could be obtained if all the brakes were functioning optimally (i.e., simultaneously on the verge of locking)? Comment on the significance of the difference. What is the front/rear brake force distribution in the optimal configuration? Does this seem like a reasonable ratio?
c) By the time the team repairs their vehicle, returning it to its design spec of front to rear brake force distribution fixed at 70%/30%, it has begun to rain, and the coefficient of friction of the surface is reduced to $\mu = 0.4$. Will the front or rear tires now lock first? What is the deceleration when the first lockup occurs?

3.6 Consider the student competition vehicle from the previous problem. If the front to rear brake force distribution fixed at 75%/25%, and the coefficient of friction of the surface is $\mu = 0.8$:

a) What is the maximum deceleration that can be obtained on level ground before a tire locks? Which tire locks first in this situation?
b) What about on a downhill slope of 20%, before a tire locks? Which tire locks first in this situation?
c) What about on an uphill slope of 20%? Which tire locks first in this situation?

4

Linear Dynamic Models

A number of very well known vehicle models have been developed over the years. Most of these are relatively small linear models that can be manipulated by hand, but provide results that allow valuable qualitative insight into vehicle behaviour. It is important to note that not all problems can be safely approached with the assumption of linear behaviour. The modeller must determine the required level of fidelity of the model and whether the behaviour is actually well described by linear relationships. Often, a simple linear model may better meet the needs of the problem at hand, but in certain cases, nonlinear effects cannot reasonably be ignored. At the same time, one should recognize that solution of a nonlinear problem will often require a significant increase in computational effort for a solution that provides only marginal improvement in accuracy.

This chapter contains a discussion of many of the classical linear dynamic models, including the yaw plane handling model, a simple truck and trailer model, and several ride quality models. Analysis of each of these models, and some comments on how one should interpret the results will follow their development. Those readers who are unfamiliar with some of the mathematical tools will find more detail provided in Chapter 7.

4.1 The Yaw Plane Model

Also widely known as the *bicycle model*, the yaw plane model has been used extensively for vehicle handling studies since its introduction, and has appeared in many variations in the literature. The nickname is applied because the effect of the width of the vehicle is considered unimportant during a certain aspect of the model development. When the vehicle is pictured in a view from above with the width neglected, its appearance is similar to a bicycle. The model itself has no connection to bicycle dynamics, although a dynamic model of a bicycle is discussed in Chapter 6. Of course, the vehicle width influences

Fundamentals of Vehicle Dynamics and Modelling: A Textbook for Engineers with Illustrations and Examples, First Edition. Bruce P. Minaker.

Companion website: www.wiley.com/go/minaker/vehicle-dynamics

the lateral weight transfer experienced while cornering, which in turn affects tire performance, but this is a secondary effect, and is ignored in the yaw plane model. Weight transfer effects and tire performance are discussed in more detail in Chapter 2.

The independent motions considered, or the *degrees of freedom* (DOF), of the yaw plane model are the lateral velocity v, and the yaw velocity r of the vehicle. The forward speed u is assumed to be under driver control, and held constant, and as a result is treated as a parameter of the model rather than a variable. The motion is assumed to take place on a flat and level road, so all other motions, e.g., roll or heave, are ignored. The constant forward speed is more precisely called a *nonholonomic* constraint; this implies that despite having only two degrees of freedom, three position coordinates are required to fully specify the state of the vehicle: the (x, y) location of the centre of mass, and the heading angle ψ. Despite using only two degrees of freedom, the model provides a great deal of insight into vehicle handling.

The mass m and yaw moment of inertia I_{zz} are the relevant inertial properties of the vehicle, while the distances from the centre of mass to the front and rear axles, denoted a and b respectively, provide the necessary geometric information. The *cornering stiffnesses* of the front and rear tires are denoted c_f and c_r, respectively. The cornering stiffness is a measure of the amount of lateral force produced by a rolling tire; Chapter 2 provides a more detailed description of tire behaviour. The lateral forces acting at each of the front and rear axles are Y_f and Y_r, respectively. The steering angle of the front tires, assumed to be the same on the left and right side, is δ_f. A schematic diagram is shown in Figure 4.1.

The model is formed by starting with the Newton–Euler sums. The velocities of the vehicle are defined in a frame attached to the vehicle, and rotating with it. When differentiating a vector that is defined in a rotating frame, the result is the derivative relative to the rotating frame, plus the effect caused by the rotation

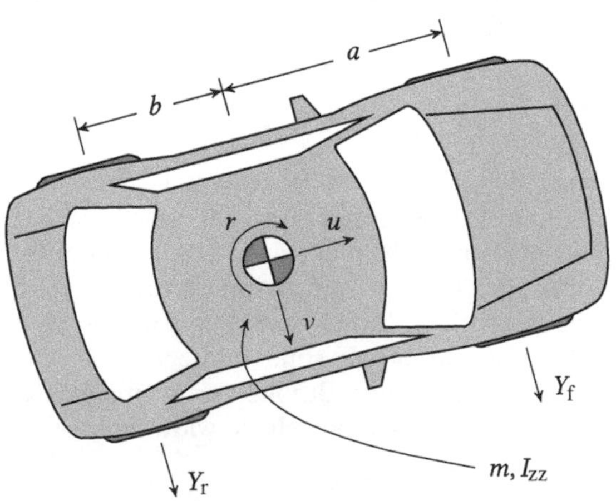

Figure 4.1 The yaw plane model treats the vehicle as a single rigid body, with motion constrained to the horizontal plane. It is commonly referred to as the *bicycle model* due to the assumption about width effects.

of the frame. In this case, the vehicle and the frame of reference have the same angular velocity. The linear velocity vector is denoted $\boldsymbol{v}$ and the angular velocity vector is denoted $\boldsymbol{\omega}$.

$$\sum f = \sum X\hat{\mathbf{i}} + Y\hat{\mathbf{j}} + Z\hat{\mathbf{k}} = m(\dot{\mathbf{v}} + \boldsymbol{\omega} \times \mathbf{v}) = m((\dot{u}\hat{\mathbf{i}} + \dot{v}\hat{\mathbf{j}}) + r\hat{\mathbf{k}} \times (u\hat{\mathbf{i}} + v\hat{\mathbf{j}})) \tag{4.1}$$

The longitudinal equation is not relevant, and ignored. The lateral force equation is found by eliminating all but the $\hat{\mathbf{j}}$ component terms, allowing the vector notation to be dropped. The result is:

$$\sum Y = Y_\mathrm{f} + Y_\mathrm{r} = m(\dot{v} + ru) \tag{4.2}$$

The general equation for rotation is:

$$\sum \boldsymbol{m}_\mathrm{G} = \sum L\hat{\mathbf{i}} + M\hat{\mathbf{j}} + N\hat{\mathbf{k}} = \mathbf{I}_\mathrm{G}\boldsymbol{\alpha} + \boldsymbol{\omega} \times \mathbf{I}_\mathrm{G}\boldsymbol{\omega} \tag{4.3}$$

As the rotation of the reference frame is confined to the vertical direction only, the moment equation simplifies considerably, and depends only on rate of change of magnitude of the angular velocity vector. Considering only the $\hat{\mathbf{k}}$ component:

$$\sum N = aY_\mathrm{f} - bY_\mathrm{r} = I_{zz}\dot{r} \tag{4.4}$$

The two equations of motion can be gathered together into a single vector equation.

$$\begin{Bmatrix} \sum Y \\ \sum N \end{Bmatrix} = \begin{bmatrix} 1 & 1 \\ a & -b \end{bmatrix} \begin{Bmatrix} Y_\mathrm{f} \\ Y_\mathrm{r} \end{Bmatrix} = \begin{Bmatrix} m(\dot{v} + ur) \\ I_{zz}\dot{r} \end{Bmatrix} \tag{4.5}$$

The dynamic equations depend on the lateral forces developed by the tires, Y_f, and Y_r. There are many factors that determine the forces generated between the tire and the ground, but the primary cause of lateral force is misalignment between the direction the tire is travelling, and the direction it is pointing. In order to maintain its simplicity, the bicycle model uses a linear tire model that treats the lateral force as proportional to the tire slip angle α only. Any longitudinal forces or self-aligning moments generated by the tire are ignored. The tire forces are then given by $Y_\mathrm{f} = -c_\mathrm{f}\alpha_\mathrm{f}$, and $Y_\mathrm{r} = -c_\mathrm{r}\alpha_\mathrm{r}$, or:

$$\begin{Bmatrix} Y_\mathrm{f} \\ Y_\mathrm{r} \end{Bmatrix} = -\begin{bmatrix} c_\mathrm{f} & 0 \\ 0 & c_\mathrm{r} \end{bmatrix} \begin{Bmatrix} \alpha_\mathrm{f} \\ \alpha_\mathrm{r} \end{Bmatrix} \tag{4.6}$$

Because the force terms include the contribution of both the left and right side of the vehicle, the properties of both front tires are lumped into an effective single central front tire, and both rear tires are lumped to an effective single central rear tire, i.e., the cornering stiffness used in the model would be double that obtained from the measurement of a single tire. Note that the sign change is used to preserve the sign convention of the model; a positive slip

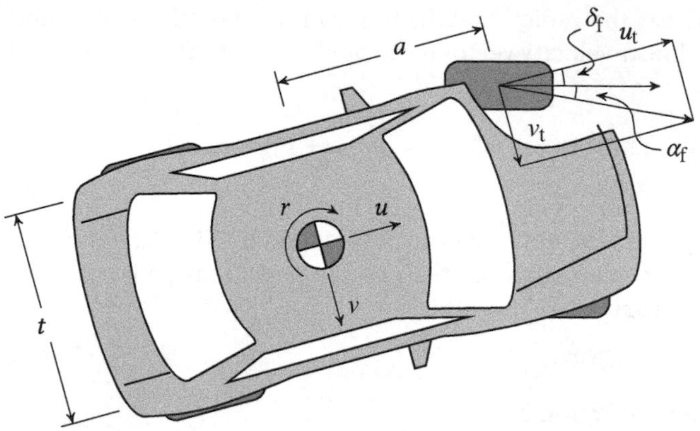

Figure 4.2 The tire slip angle is the difference between the direction the tire is pointing, and the direction it is travelling. It can be determined from the kinematics of the vehicle.

angle generates a resistive negative tire force. The slip angles are calculated from geometry, knowing that the velocity of the tire is defined by the velocity of the vehicle through kinematic relationships.

Consider the left front tire as an example, as shown in Figure 4.2. Using a relative velocity kinematics approach, the velocity of the tire contact point can be found as:

$$\mathbf{v}_t = \mathbf{v}_G + \mathbf{v}_{t/G} = u\hat{\mathbf{i}} + v\hat{\mathbf{j}} + r\hat{\mathbf{k}} \times (a\hat{\mathbf{i}} - \frac{t}{2}\hat{\mathbf{j}}) \tag{4.7}$$

where $\frac{t}{2}$ is half the track width. The tire's lateral speed will be the sum of the lateral velocity of the vehicle, plus the effect of the yaw velocity: $v_t = v + ra$. A similar result follows for the forward speed: $u_t = u + \frac{rt}{2}$ The slip angle, the steer angle, and the velocities are then related by the following expression:

$$\tan(\alpha_f + \delta_f) = \frac{v + ra}{u + \dfrac{rt}{2}} \tag{4.8}$$

For the range of vehicle speeds where the model is applicable, $u \gg \frac{rt}{2}$, so $u_t \approx u$. The width of the vehicle is ignored in the calculation of the longitudinal tire speed, and the result is then applicable to the tires on either the left or right side, as the resulting slip angle is the same. For small slip angles, the following approximations are adequate:

$$\alpha_f + \delta_f = \frac{v + ra}{u} \tag{4.9}$$

$$\alpha_r = \frac{v - rb}{u} \tag{4.10}$$

or:

$$\begin{Bmatrix} \alpha_f \\ \alpha_r \end{Bmatrix} = \frac{1}{u} \begin{bmatrix} 1 & a \\ 1 & -b \end{bmatrix} \begin{Bmatrix} v \\ r \end{Bmatrix} - \begin{Bmatrix} \delta_f \\ 0 \end{Bmatrix} \tag{4.11}$$

Equation (4.11) can be substituted into Equation (4.6), which can in turn be substituted into Equation (4.5) to give a complete first order model.

$$\begin{bmatrix} m & 0 \\ 0 & I_{zz} \end{bmatrix} \begin{Bmatrix} \dot{v} \\ \dot{r} \end{Bmatrix} + \frac{1}{u} \begin{bmatrix} c_f + c_r & ac_f - bc_r + mu^2 \\ ac_f - bc_r & a^2 c_f + b^2 c_r \end{bmatrix} \begin{Bmatrix} v \\ r \end{Bmatrix} = \begin{bmatrix} c_f \\ ac_f \end{bmatrix} \{\delta_f\} \tag{4.12}$$

In a simplified notation:

$$\mathbf{M}\dot{\boldsymbol{x}} + \mathbf{L}\boldsymbol{x} = \mathbf{F}\delta_f \tag{4.13}$$

The resulting equation is well known, and has been used in many studies in the literature over the years. As mentioned, it ignores certain factors in the interest of developing a usable linear model, but has proven to be reasonably accurate, and to predict the fundamental dynamic characteristics of a typical passenger car. The model will next be manipulated to focus on several areas of interest, beginning with the steady state solution.

4.1.1 Steady State Analysis

By assuming a steady state motion, where the vehicle is cornering at a constant rate and the lateral and yaw velocities have reached constant values, i.e., $\dot{\boldsymbol{x}} = \mathbf{0}$, the differential equations of motion simplify to a pair of algebraic equations.

$$\frac{1}{u} \begin{bmatrix} c_f + c_r & ac_f - bc_r + mu^2 \\ ac_f - bc_r & a^2 c_f + b^2 c_r \end{bmatrix} \begin{Bmatrix} v \\ r \end{Bmatrix} = \begin{bmatrix} c_f \\ ac_f \end{bmatrix} \{\delta_f\} \tag{4.14}$$

An inversion of the $\mathbf{L}$ matrix allows the equations to be solved for both the steady state lateral velocity and yaw rate, per unit of steer angle, v/δ_f and r/δ_f; these quantities are often referred to as the *steady state gains.*

$$\begin{Bmatrix} v/\delta_f \\ r/\delta_f \end{Bmatrix} = u \begin{bmatrix} c_f + c_r & ac_f - bc_r + mu^2 \\ ac_f - bc_r & a^2 c_f + b^2 c_r \end{bmatrix}^{-1} \begin{bmatrix} c_f \\ ac_f \end{bmatrix}$$

$$= \frac{u \begin{bmatrix} a^2 c_f + b^2 c_r & -ac_f + bc_r - mu^2 \\ -ac_f + bc_r & c_f + c_r \end{bmatrix} \begin{bmatrix} c_f \\ ac_f \end{bmatrix}}{(c_f + c_r)(a^2 c_f + b^2 c_r) - (ac_f - bc_r + mu^2)(ac_f - bc_r)} \tag{4.15}$$

Yaw Rate

The steady state yaw rate gain becomes:

$$\frac{r}{\delta_f} = \frac{u}{a + b - \dfrac{mu^2(ac_f - bc_r)}{(a+b)c_f c_r}} \tag{4.16}$$

This equation can be manipulated to help give a fundamental understanding of cornering behaviour. First, the cornering radius R of the path of the vehicle is found from:

$$u = rR \tag{4.17}$$

This simple expression can be confusing at first, but can be clarified when compared to the familiar angular velocity relation $v = \omega r$, (with careful attention to the changes in notation). Next, the *kinematic cornering radius* R_0 is defined. If the assumption of straight ahead motion at both axles is used, this result falls from simple geometry, assuming a small steer angle, such that $\tan \delta_f \approx \delta_f$. This is the cornering radius the vehicle will follow when cornering at very low speed, where the lateral forces are low, and tire slip angle is negligible.

$$R_0 = \frac{a+b}{\delta_f} \tag{4.18}$$

A combination of these three results gives:

$$\frac{R}{R_0} = 1 - \frac{mu^2(ac_f - bc_r)}{(a+b)^2 c_f c_r} \tag{4.19}$$

From inspection of this equation, it is apparent that the term $ac_f - bc_r$ becomes important in determining the response of the vehicle. If $ac_f < bc_r$, as would likely be the case if the centre of mass is towards the front of the vehicle, then the cornering radius will increase with increasing forward velocity. If a driver attempts to follow a fixed circular path, but slowly increases his speed while doing so, the vehicle will tend to track outside the path, unless the steer angle is increased to compensate. This behaviour is termed *understeer*. If $ac_f > bc_r$, as would likely be the case for a rear heavy vehicle, then the cornering radius will tend to decrease with increasing velocity, so less steer input is required to maintain a fixed radius path. The vehicle is said to *oversteer*. Finally, if $ac_f = bc_r$, the cornering radius is always R_0, independent of forward velocity, i.e., changing the vehicle speed on a fixed radius path requires no change in steer input from the driver to maintain that path. The vehicle is said to *neutral steer*. In practice, despite the apparent advantage of a neutral steering or *balanced* vehicle design, almost all passenger cars will exhibit understeer behaviour, although this is may be influenced by a number of factors that have not been considered in this model.

Plots of the yaw rate gain, Equation (4.16), are shown in Figure 4.3, also using the values given in Table 4.1. The yaw rate gain is very different for oversteering and understeering vehicles. For understeering vehicles, the gain increases with increasing speed to a maximum value, before dropping off slightly to a near constant value. The speed at which the maximum gain occurs can be found by

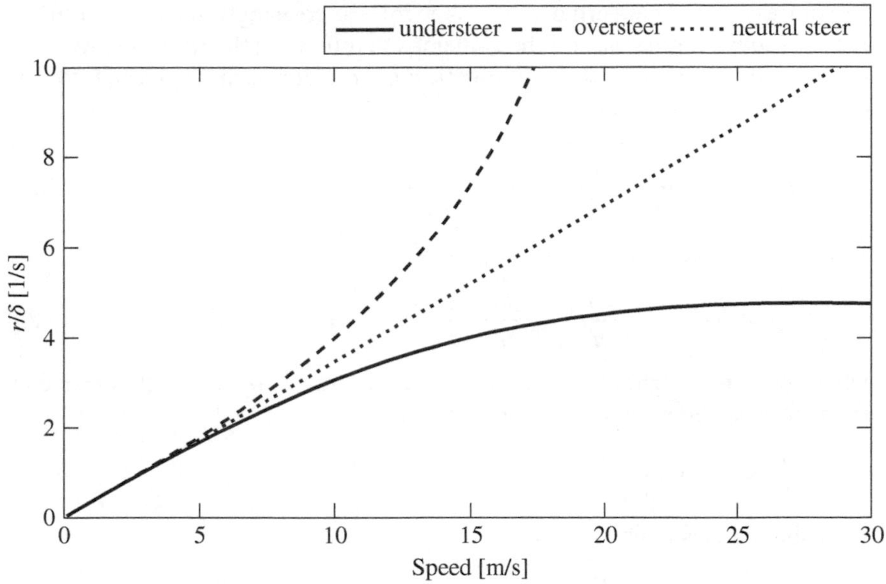

Figure 4.3 The steady state yaw rate gain r/δ is a function of speed. Note that the oversteering vehicle becomes increasing sensitive near the critical speed u_{crit}, while the understeering vehicle reaches a maximum at the characteristic speed u_{char}.

Table 4.1 Yaw plane model parameter values

Quantity	Notation	Value
front axle dimension	a	1.189 m
rear axle dimension	b	1.696 m
mass	m	1 730 kg
yaw inertia	I_{zz}	3 508 kg m^2
front tire cornering stiffness	c_f	80 000 N m/rad
rear tire cornering stiffness	c_r	80 000 N m/rad

Notes:

- the default values give the understeer configuration, the values of a and b are reversed to produce the oversteer configuration
- the value of the cornering stiffness is a sum of the left and right side tire for each axle

setting the derivative to zero, and is termed the *characteristic speed*. It is given as:

$$u_{char} = \sqrt{\frac{(a+b)^2 c_f c_r}{m(bc_r - ac_f)}} = \sqrt{\frac{g(a+b)}{k_{us}}} \tag{4.20}$$

In contrast, the oversteering vehicle becomes increasingly sensitive to driver input as speed increases, and the gain goes to infinity at the *critical speed*. Setting the denominator in the gain function to zero allows an expression for the critical speed to be found.

$$u_{\text{crit}} = \sqrt{\frac{(a+b)^2 c_{\text{f}} c_{\text{r}}}{m(ac_{\text{f}} - bc_{\text{r}})}} = \sqrt{\frac{g(a+b)}{-k_{\text{us}}}} \tag{4.21}$$

Defining the *understeer gradient* k_{us} as:

$$k_{\text{us}} = \frac{Z_{\text{f}}}{c_{\text{f}}} - \frac{Z_{\text{r}}}{c_{\text{r}}} = \frac{mg}{a+b}\left(\frac{b}{c_{\text{f}}} - \frac{a}{c_{\text{r}}}\right) \tag{4.22}$$

where Z_{f} is the weight carried by the front axle, and Z_{r} is the weight carried by the rear axle, allows Equation (4.19) to be rewritten as:

$$\frac{R}{R_0} = 1 + k_{\text{us}}\frac{u^2}{g(a+b)} \tag{4.23}$$

Note that this also implies that:

$$\delta_{\text{f}} = \frac{a+b}{R} + k_{\text{us}}\frac{u^2}{gR} \tag{4.24}$$

A close examination of Equation (4.24) is revealing. It illustrates how the steer angle changes with speed. The first term is a fixed amount required by kinematics, and the second term adds on the amount of extra steer angle required to accommodate tire slip. A positive understeer gradient indicates that the vehicle is understeering. Examination of the second term reveals that the understeer gradient is multiplied by the lateral acceleration, with division by the gravitational constant allowing it to be expressed in units of steer per 'g' of lateral acceleration. Plots of Equation (4.24) are given for both an understeering and oversteering vehicle in Figures 4.4 and 4.5, using values from Table 4.1. The values are chosen to represent a typical passenger minivan. As one might expect, the understeering vehicle needs increasing steer input with increasing speed and decreasing cornering radius. The oversteering vehicle also requires increasing steer angle with decreasing radius, but decreasing steer angle with increasing speed.

If one was to exceed the critical speed in an oversteering vehicle, a strange result occurs: the yaw rate gain changes sign, as does the steer angle. In this situation, termed *countersteer*, a steer angle to the left is required for the vehicle to maintain a corner to the right, and vice versa. While it seems counterintuitive, a careful study of many dirt-track style racing vehicles will confirm that this situation occurs quite regularly, and is more commonly known as *drifting*. Note that in the case of racing vehicles, the value of c_{r} can be significantly reduced by the large longitudinal forces acting at the tire, lowering the critical speed. It is also important to remember that these expressions only govern the steady

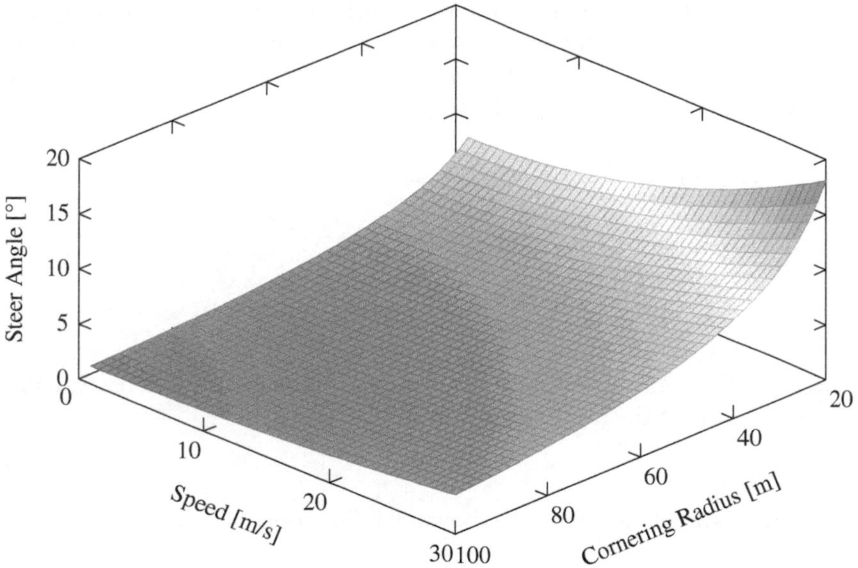

Figure 4.4 The steady state steer angle, shown in the understeer configuration, is a function of forward speed and cornering radius. The steer angle required increases with increasing speed and with decreasing radius.

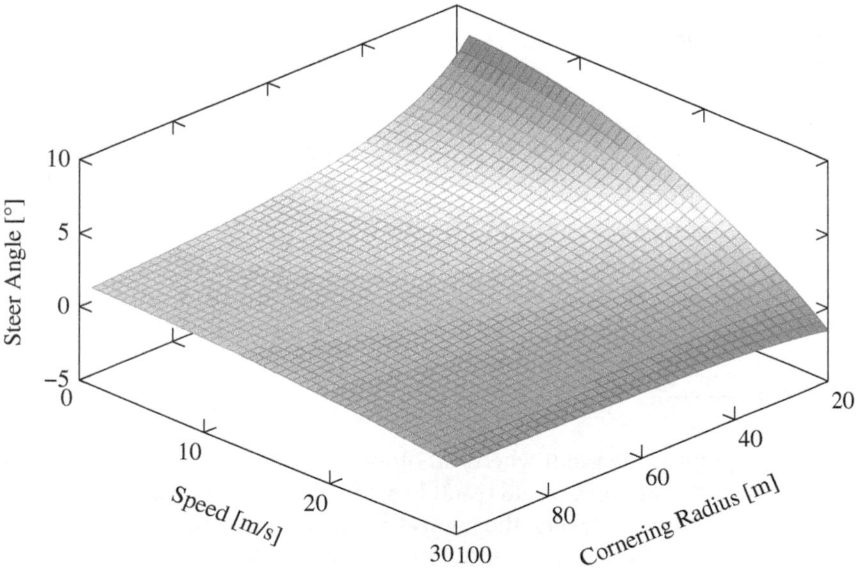

Figure 4.5 The steady state steer angle in the oversteer configuration decreases with increasing speed and increasing radius. At the critical speed, the steer angle crosses zero.

state behaviour; the transient condition during corner entry will require a steer input in the direction of the turn to initiate the corner. Only once the vehicle has settled into a steady turn is the steer angle reversed. Making use of these definitions, for an oversteering vehicle, one can also write:

$$\frac{R}{R_0} = 1 - \left(\frac{u}{u_{\text{crit}}}\right)^2 \tag{4.25}$$

or for an understeering vehicle:

$$\frac{R}{R_0} = 1 + \left(\frac{u}{u_{\text{char}}}\right)^2 \tag{4.26}$$

A measure of understeer is the *understeer angle*, defined as the difference in tire slip angles between the front and rear axle.[1]

$$\alpha_{\text{u}} = \alpha_{\text{r}} - \alpha_{\text{f}} \tag{4.27}$$

With some manipulation, it can be shown that Equation (4.24) can be rewritten as:

$$\delta_{\text{f}} = \frac{a+b}{R} + \alpha_{\text{u}} \tag{4.28}$$

which implies that:

$$\alpha_{\text{u}} = k_{\text{us}}\frac{u^2}{gR} \tag{4.29}$$

or:

$$\alpha_{\text{u}} = \delta\left(\frac{\left(\frac{u}{u_{\text{char}}}\right)^2}{1+\left(\frac{u}{u_{\text{char}}}\right)^2}\right) \tag{4.30}$$

An alternate means of characterizing the understeer properties of a vehicle is to define the *neutral point*, a distance *d* behind the front axle, where:

$$d = \frac{(a+b)c_{\text{r}}}{c_{\text{f}} + c_{\text{r}}} \tag{4.31}$$

The neutral point is the point where, all other things being equal, locating the centre of mass at that point would result in a neutral steering vehicle. The *static margin* is the distance between the neutral point and the centre of mass. If the centre of mass is in front of the neutral point, the vehicle will understeer.

1 Some authors use the alternate definition of $\alpha_{\text{f}} - \alpha_{\text{r}}$ due to varying sign convention of the tire slip angle.

Body Slip Angle

When a vehicle is cornering, it will not always point exactly in the direction of travel. The *body slip angle* β can be approximated, for small angles, as the ratio of lateral sliding speed to forward speed.

$$\beta = \frac{v}{u} \tag{4.32}$$

As the vehicle enters a corner, there will be some transient phase, where the body slip angle will fluctuate, followed by a steady state phase, where the slip angle will approach some fixed value as the vehicle corners. Note that as a result of the definition of body slip angle, the lateral acceleration can be expressed as $\dot{v} + ur = u(\dot{\beta} + r)$. The steady state β to δ_f ratio, or steady state body slip angle gain, can be found from the steady state sliding speed gain.

$$\frac{\beta}{\delta_f} = \frac{b - \dfrac{amu^2}{(a+b)c_r}}{a + b - \dfrac{mu^2(ac_f - bc_r)}{(a+b)c_f c_r}} \tag{4.33}$$

An examination of the resulting expression reveals more interesting distinctions between an understeering and an oversteering vehicle. Note that the sign of the body slip angle gain can be related to a simple physical condition. When the vehicle is cornering, if the rear axle of the vehicle is tracking on a path inside of the front axle, then the gain is positive. If the rear axle tracks outside the front, then the gain is negative. A common misunderstanding is to equate tail-out behaviour with the oversteer condition, but this is not entirely correct. A plot of the expression, shown in Figure 4.6, indicates that understeering and oversteering vehicles behave similarly at low speed, where the rear axle tracks inside the front for both configurations. The reader may recall seeing a set of tracks left behind when a driver has executed a U-turn on a dirt- or snow-covered surface, perhaps noticing their relative positioning. Both understeering and oversteering vehicles will transition to a tail-out condition as speed increases, but this transition occurs at a lower speed for the oversteering vehicle, which will typically have higher body slip angle gains in general. The expression also shows that the understeering car approaches a limit in slip angle gain, as speed increases. The limit can be found as:

$$\left.\frac{\beta}{\delta_f}\right|_{\text{lim}} = \frac{ac_f}{ac_f - bc_r} \tag{4.34}$$

Once in the tail-out condition, the oversteering car has a body slip angle gain that increases with increasing speed, and goes to infinity asymptotically at the critical speed. Similar to the yaw rate gain, the body slip angle gain also changes sign above the critical speed. Note that if the speed is slowly increased

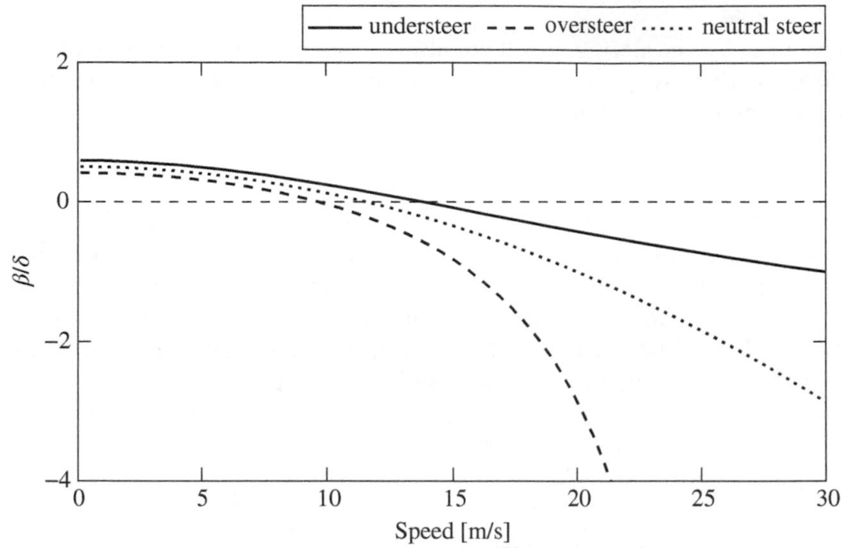

Figure 4.6 The steady state body slip angle gain β/δ is a function of speed. Note that the oversteering configuration has an asymptote at $u = u_{crit}$, and transitions to a negative value (tail-out orientation) at a lower speed than the understeer configuration.

through the critical speed in a steady corner, it is generally the steering angle that changes sign, and not the yaw rate or body slip angle. The change in sign of the steer angle coincides with the change in sign of the gains. Using the values of the steer angle required to navigate a corner of fixed radius in a steady state, the steady state slip angle can also be expressed as:

$$\beta = \frac{b}{R} - \frac{amu^2}{(a+b)c_r R} = \frac{b}{R} - \frac{Z_r}{c_r}\frac{u^2}{gR} \tag{4.35}$$

The expression can be recognized as similar in form to Equation (4.24). The steady state body slip angle has a constant term based only on geometry, plus another term that relates to the lateral acceleration.

4.1.2 Transient Analysis

While very useful, the steady state analysis does not tell the entire story about the yaw plane vehicle. As the vehicle enters or exits a turn, there is a transition from straight ahead running to a curved path, or vise versa. During the transition, there is some transient motion that occurs, and the previous steady state results do not apply. To conduct a transient analysis, it is assumed that the vehicle has been slightly disturbed from straight ahead running, but that the driver does not react, and holds the steer input fixed at zero. If one assumes that the sliding speed and yaw rate can be described by an exponential function following the disturbance, a simple but interesting result can be found. In fact, a study of linear differential equations shows that only in special cases is the transient solution something other than an exponential or a sinusoidal curve, so it is not so big of an assumption.

$$\boldsymbol{x}(t) = \boldsymbol{x}_0 \mathrm{e}^{st} \tag{4.36}$$

Differentiation gives:

$$\dot{\boldsymbol{x}}(t) = \boldsymbol{x}_0 s \mathrm{e}^{st} \tag{4.37}$$

Substitution into the equation of motion gives:

$$\mathbf{M}\boldsymbol{x}_0 s\mathrm{e}^{st} + \mathbf{L}\boldsymbol{x}_0 \mathrm{e}^{st} = \mathbf{0} \tag{4.38}$$

or:

$$[\mathbf{M}s + \mathbf{L}]\boldsymbol{x}_0 = \mathbf{0} \tag{4.39}$$

A careful inspection will reveal that this is an eigenvector problem. See Chapter 7 for an expanded discussion on eigenvector problems. An expansion of the determinant above will give the characteristic equation of the problem. The roots of the characteristic equation are the eigenvalues, and they form the foundation of the transient solution. The characteristic equation of the yaw plane model is expanded below.

$$\det[\mathbf{M}s + \mathbf{L}] = 0 \tag{4.40}$$

$$\det\begin{bmatrix} ms + \dfrac{c_\mathrm{f} + c_\mathrm{r}}{u} & \dfrac{ac_\mathrm{f} - bc_\mathrm{r}}{u} + mu \\ \dfrac{ac_\mathrm{f} - bc_\mathrm{r}}{u} & I_{zz}s + \dfrac{a^2c_\mathrm{f} + b^2c_\mathrm{r}}{u} \end{bmatrix} = 0 \tag{4.41}$$

$$\left(ms + \frac{c_\mathrm{f} + c_\mathrm{r}}{u}\right)\left(I_{zz}s + \frac{a^2c_\mathrm{f} + b^2c_\mathrm{r}}{u}\right) - \left(\frac{ac_\mathrm{f} - bc_\mathrm{r}}{u} + mu\right)\left(\frac{ac_\mathrm{f} - bc_\mathrm{r}}{u}\right) = 0 \tag{4.42}$$

$$s^2 + \left(\frac{a^2c_\mathrm{f} + b^2c_\mathrm{r}}{I_{zz}u} + \frac{c_\mathrm{f} + c_\mathrm{r}}{mu}\right)s + \frac{(a+b)^2c_\mathrm{f}c_\mathrm{r}}{mI_{zz}u^2} - \frac{ac_\mathrm{f} - bc_\mathrm{r}}{I_{zz}} = 0 \tag{4.43}$$

In this case, because the system is of dimension two, the characteristic equation is a quadratic, meaning the eigenvalues can be found directly from:

$$s = -\frac{d_1}{2} \pm \sqrt{\left(\frac{d_1}{2}\right)^2 - d_0} \tag{4.44}$$

where:

$$d_0 = \frac{(a+b)^2 c_f c_r}{m I_{zz} u^2} - \frac{a c_f - b c_r}{I_{zz}} \tag{4.45}$$

$$d_1 = \frac{a^2 c_f + b^2 c_r}{I_{zz} u} + \frac{c_f + c_r}{m u} \tag{4.46}$$

By a careful study of the characteristic equation, one can determine the eigenvalues, and perhaps more importantly, their signs. The signs of the real parts of the eigenvalues determine whether the transient solution is shrinking or growing with time, i.e., whether the system is stable or unstable. In this case, because the system is only of dimension two, the values of the roots can be found directly. For systems of larger dimension, numerical methods must be used to find the eigenvalues. In many cases, it is not necessary to compute the eigenvalues, as the Routh–Hurwitz criterion can provide information on the stability, without finding the actual eigenvalues. The Routh–Hurwitz criterion states that a quadratic will not have unstable roots when all the coefficients have the same sign. There are additional criteria for higher order systems.

The coefficient of the linear term in the characteristic equation is always positive in this model (assuming forward speed), so eigenvalues with positive real parts only occur when the constant term becomes negative.

$$\frac{(a+b)^2 c_f c_r}{m I_{zz} u^2} - \frac{a c_f - b c_r}{I_{zz}} < 0 \tag{4.47}$$

An examination of the expression shows that the first term is always positive, and the second is negative if $a c_f < b c_r$. As a result, the difference must always be positive, i.e., an understeering vehicle is stable at any forward speed. This result explains the popularity of understeering handling characteristics with most automotive manufacturers. If the vehicle is oversteering, then the condition for instability is:

$$u > \sqrt{\frac{(a+b)^2 c_f c_r}{m(a c_f - b c_r)}} = u_{crit} \tag{4.48}$$

The unstable root only occurs when an oversteering vehicle is operating above its critical speed. Not only does the direction of the steady component of the required steer input reverse when operating an oversteering vehicle above its critical speed, but the vehicle will also require active correction from the driver to counteract its tendency to develop large yawing and sliding motions. The roots of the characteristic equation are plotted as a function of forward speed in Figures 4.7 and 4.8. The roots of both the understeering and oversteering vehicles are both real and very negative at low speed, indicating a very stable system. The oversteering vehicle has two real roots, one of which will become positive when it reaches the critical speed, indicating low speed stability, and high speed instability.

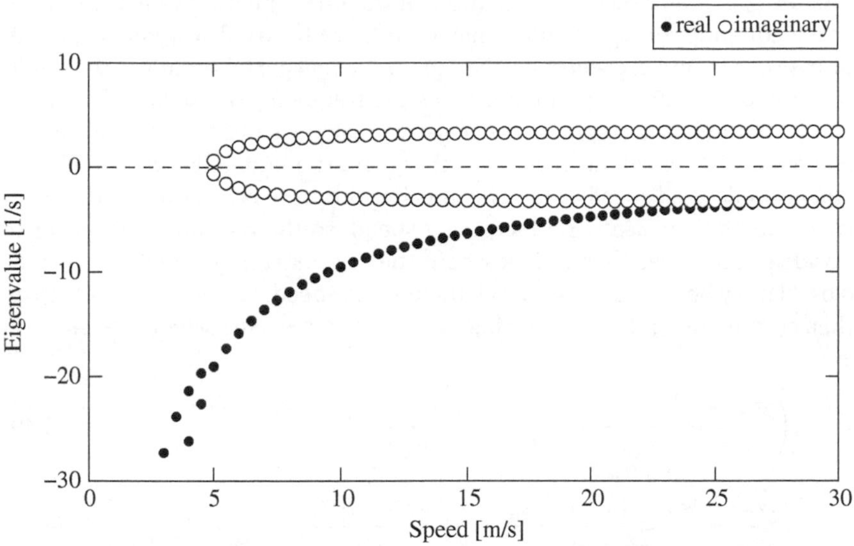

Figure 4.7 When the vehicle is in the understeer configuration and above the transition speed, the eigenvalues of the equation of motion are complex with negative real parts, indicating a stable but oscillatory system. The imaginary part of the root, and therefore the associated frequency, is relatively independent of speed, while the real part indicates decreasing yaw damping as speed increases.

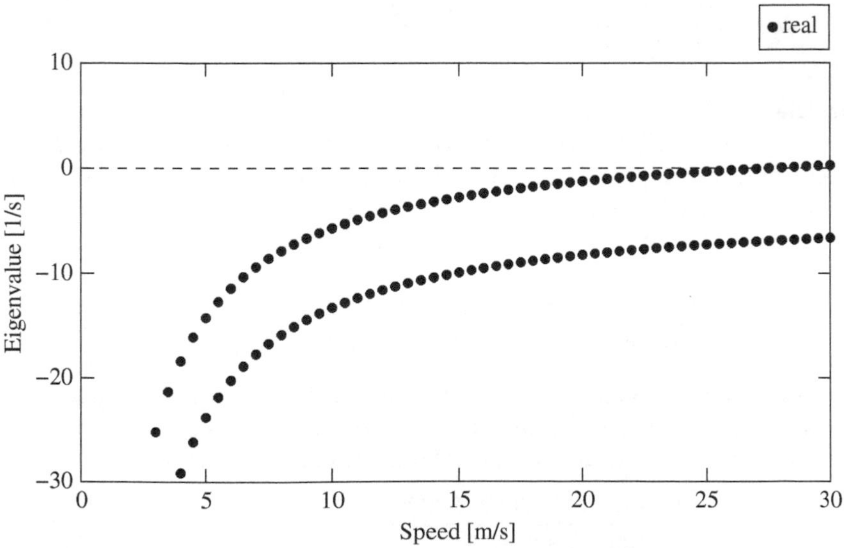

Figure 4.8 With the vehicle in the oversteer configuration, the eigenvalues of the equation of motion remain real, but one becomes postive above the critical speed, indicating an unstable system. However, the unstable root is relatively small, implying slow exponential growth.

As the speed increases, the real roots of the understeering vehicle merge to form a complex conjugate, indicating a tendency for oscillatory behaviour. In the case of a complex eigenvalue, the sign of the real part determines the stability, while the imaginary part determines the frequency of oscillation. Despite the oscillation in the response, the negative real parts indicate that the amplitude of oscillation will decrease with time, and the understeering vehicle will remain stable at all speeds. However, it is important to note also that the effective damping decreases with increasing speed, so the motion takes longer to subside at high speed. The point where the understeering vehicle transitions to oscillatory behaviour is not its characteristic speed, as one might anticipate when comparing to the oversteering case. Rather, it occurs when $d_1^2 - 4d_0 = 0$, or:

$$\left(\frac{a^2c_f + b^2c_r}{I_{zz}u} + \frac{c_f + c_r}{mu}\right)^2 - 4\frac{(a+b)^2c_fc_r}{mI_{zz}u^2} + 4\frac{ac_f - bc_r}{I_{zz}} = 0 \tag{4.49}$$

$$u^2\frac{ac_f - bc_r}{I_{zz}} = \frac{(a+b)^2c_fc_r}{mI_{zz}} - \frac{1}{4}\left(\frac{a^2c_f + b^2c_r}{I_{zz}} + \frac{c_f + c_r}{m}\right)^2 \tag{4.50}$$

Assuming that $ac_f - bc_r \neq 0$

$$u_{\text{trans}}^2 = \frac{(a+b)^2c_fc_r}{m(ac_f - bc_r)} - \frac{I_{zz}}{4(ac_f - bc_r)}\left(\frac{a^2c_f + b^2c_r}{I_{zz}} + \frac{c_f + c_r}{m}\right)^2 \tag{4.51}$$

$$u_{\text{trans}}^2 = \frac{I_{zz}}{4(bc_r - ac_f)}\left(\frac{a^2c_f + b^2c_r}{I_{zz}} + \frac{c_f + c_r}{m}\right)^2 - u_{\text{char}}^2 \tag{4.52}$$

Example

Using the values in Table 4.1, compute the characteristic speed, the natural frequency and damping ratio at the critical speed, and the speed where the motion transitions to oscillatory.

The characteristic speed can be computed directly from Equation (4.20).

$$u_{\text{char}} = \sqrt{\frac{(1.189 + 1.696)^2(80000)(80000)}{1730(1.696(80000) - 1.189(80000))}} = 27.6\ \text{m/s} = 99.2\ \text{km/h}$$

The result can be compared to the plots shown in Figure 4.3. The peak yaw rate sensitivity occurs at the characteristic speed. The natural frequency and damping ratio can be computed using Equations (4.44) – (4.46)

$$d_0 = \frac{(1.189 + 1.696)^2(80000)(80000)}{1730(3508)27.553^2} - \frac{1.189(80000) - 1.696(80000)}{3508} = 23.124$$

$$d_1 = \frac{1.189^2(80000) + 1.696^2(80000)}{3508(27.553)} + \frac{80000 + 80000}{1730(27.553)} = 6.9075$$

$$s = -\frac{6.9075}{2} \pm \sqrt{\left(\frac{6.9075}{2}\right)^2 - 23.124} = -3.4537 \pm 3.3460i$$

The result can be compared to the plots in Figure 4.7. From the magnitude of the eigenvalue, the natural frequency can be found.

$$\omega_n = \sqrt{-3.4537^2 + 3.3460^2} = 4.8087 \text{ rad/s} = 0.765 \text{ Hz}$$

The damping ratio is also found from the eigenvalue.

$$\zeta = \frac{3.4537}{4.8087} = 0.718$$

From Equation (4.52), the transition speed can be computed.

$$u_{\text{trans}}^2 = \frac{3508}{4(1.696 - 1.189)80000}\left(\frac{(1.189^2 + 1.696^2)80000}{3508} + \frac{160000}{1730}\right)^2 - 27.553^2$$

$$u_{\text{trans}} = 4.9037 \text{ m/s} = 17.7 \text{ km/h}$$

The result can again be compared to the plots in Figure 4.7.

Centre of Yaw

From the associated eigenvector problem, a centre of yaw rotation can be defined. If the combined yaw and lateral speed is treated as though it occurs due to a rotation around a point with no lateral speed, then the distance l from the point to the centre of mass can be found from:

$$v = -lr \tag{4.53}$$

The sign of the expression is chosen such that a positive value of l represents a point ahead of the vehicle, and a negative value represents a point behind the vehicle, as shown in Figure 4.9.

The eigenvector provides entries for each of the lateral and yaw speeds, and from their ratio, values of l can be computed. Each mode provides a yaw centre, and the total motion is a sum of the motions around each of the yaw centres. The closer a yaw centre is to the mass centre, the more yaw motion is involved in that mode; as the yaw centre moves away, the amount of lateral motion is increased. Plots of the yaw centre locations for understeering and oversteering vehicles can be seen in Figures 4.10 and 4.11 respectively.

An examination of the results for the understeer configuration show that at low speed, the centres are approximately equally spaced in front of and behind the vehicle, near the axles, although the exact distances depend strongly on the

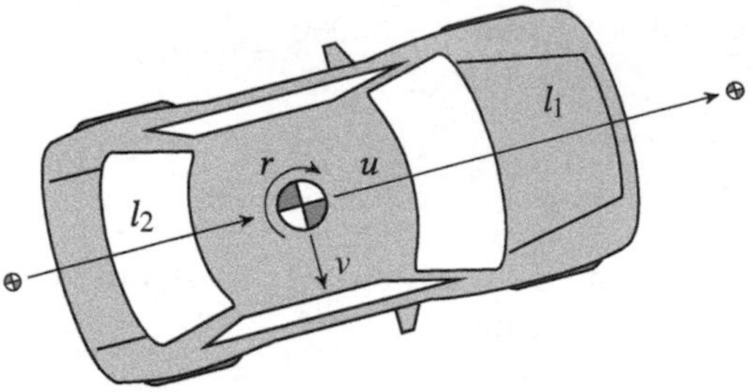

Figure 4.9 The yaw plane model has two yaw centres that can be found from the eigenvectors. In an oversteering vehicle, the lateral and yaw motions will always be in phase, so two real centres of rotation will exist. The total motion will consist of the sum of the two motions around each yaw centre. Once an understeering vehicle is above the transition speed, the yaw and lateral motions are out of phase, so no real yaw centre will exist.

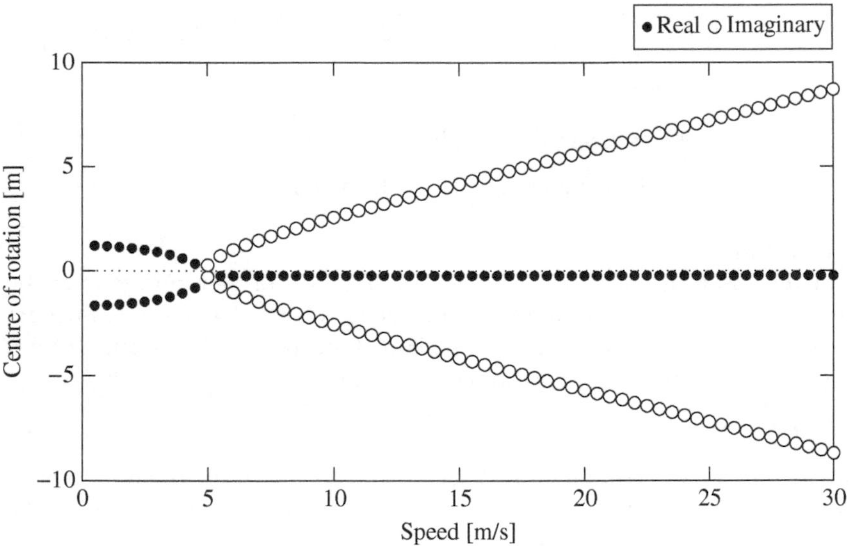

Figure 4.10 The yaw centres can be found from the eigenvectors. For an understeering vehicle, at low speed, the centres are distinct and close to the axles, but as the speed increases, they converge to a common point near the mass centre. Above the transition speed when the motion becomes oscillatory, the yaw and lateral motions are out of phase, and so the distance to the yaw centre becomes a complex value.

yaw moment of inertia. As the speed increases, the centres converge, indicating that the modes are consisting of increasing amounts of yaw motion. Above the transition speed, where the eigenvalues are complex, indicating an oscillatory motion, the distance l also becomes complex. This indicates that the lateral and yaw motions are sinusoidal and at matching frequencies, but are not occurring

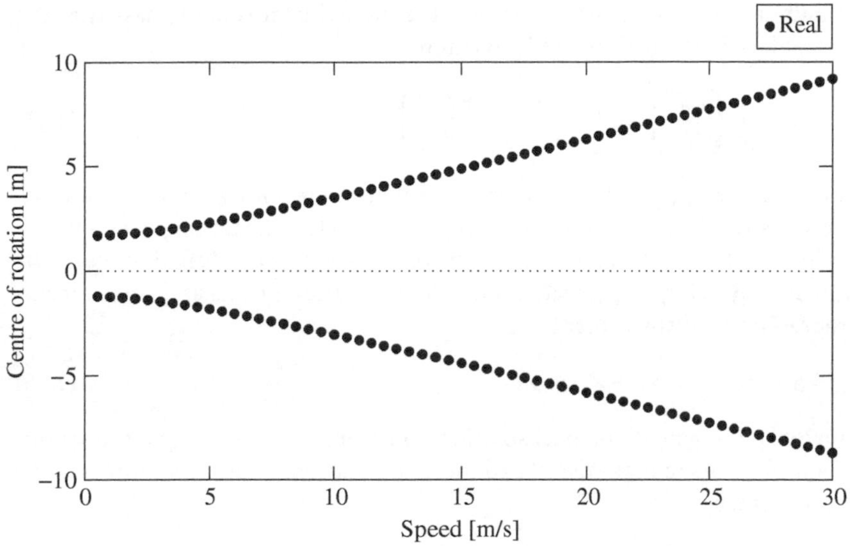

Figure 4.11 The yaw centres for an oversteering vehicle show very different behaviour. Both distances remain real valued, and the centres move farther from the vehicle as speed increases, indicating a tendency toward increased lateral motion, and decreased yaw motion in the individual modes.

in phase, i.e., the maximum yaw rate does not occur simultaneously with the maximum lateral speed.

The motions for the oversteering vehicle show that at low speed, the behaviour is quite similar to the understeering vehicle. However, as speed increases, both values of l remain real, and the centres move progressively further away from the centre of mass. The more distant yaw centres indicate that less yaw and more lateral motion is associated with each mode. Although it is not apparent from the plots, in this case, it is rotation about the yaw centre in front of the vehicle that is associated with the unstable mode.

4.1.3 Frequency Response

Another analysis that can be conducted with the yaw plane model is the frequency response. In this case, the particular solution to the equation of motion is found, and the transient solution is ignored. Both the input and the output of the model are assumed to be sinusoidal, i.e., the vehicle responds to a sinusoidal steering input with a sinusoidal lateral speed and yaw rate. The frequency response represents the sensitivity of system, i.e., the ratio of outputs to input, as a function of frequency. Assuming the input is:

$$\delta(t) = \delta_0 \cos \omega t \tag{4.54}$$

The outputs have the same frequency as the input, but are out of phase with the input and each other. They can be written:

$$\boldsymbol{x}(t) = \begin{Bmatrix} v(t) \\ r(t) \end{Bmatrix} = \begin{Bmatrix} v_0 \cos(\omega t + \phi_1) \\ r_0 \cos(\omega t + \phi_2) \end{Bmatrix} \tag{4.55}$$

where in general, $\phi_1 \neq \phi_2 \neq 0$. Assuming that the resulting motions continue to be sinusoidal, but not necessarily in phase with the excitation poses a mathematical challenge. The phase angle offsets in the sinusoidal functions could be eliminated by using an equivalent linear combination of sine and cosine terms to represent the displacement.

$$\boldsymbol{x}(t) = \boldsymbol{x}_0 \cos \omega t + \boldsymbol{x}'_0 \sin \omega t \tag{4.56}$$

From this point a mathematical solution could proceed. However, a more popular alternative is to make use of exponential notation. Using this approach, the input becomes:

$$\delta(t) = \delta_0 \mathrm{e}^{st} \tag{4.57}$$

where the substitution $s = \mathrm{i}\omega$ is made.

$$\delta(t) = \delta_0 \mathrm{e}^{\mathrm{i}\omega t} = \delta_0(\cos \omega t + \mathrm{i} \sin \omega t) \tag{4.58}$$

A similar choice is made for the outputs.

$$\boldsymbol{x}(t) = \boldsymbol{x}_0 \mathrm{e}^{st} = \boldsymbol{x}_0 \mathrm{e}^{\mathrm{i}\omega t} \tag{4.59}$$

The assumed input and outputs are substituted into the yaw plane model, Equation (4.13).

$$\mathbf{M}\boldsymbol{x}_0 \mathrm{i}\omega \mathrm{e}^{\mathrm{i}\omega t} + \mathbf{L}\boldsymbol{x}_0 \mathrm{e}^{\mathrm{i}\omega t} = \mathbf{F}\ \delta_0 \mathrm{e}^{\mathrm{i}\omega t} \tag{4.60}$$

$$\frac{\boldsymbol{x}_0}{\delta_0} = [\mathbf{M}\mathrm{i}\omega + \mathbf{L}]^{-1}\mathbf{F} \tag{4.61}$$

When using exponential notation, the linearity of the system is key, as it allows superposition of solutions. In this case, a second input of identical frequency and amplitude, but offset by 90 deg in the complex plane, is superimposed on the original input. This second imaginary input will generate a second superimposed output. However, the real part of the input will generate only the real part of the output; similarly, the imaginary part of the input generates only the imaginary part of the output. As a consequence, one may safely assume that the input is complex, and simply ignore the imaginary part of the response. The response to the real input is found by generating both solutions, and then discarding the imaginary one. See Chapter 7 for more discussion on the frequency response, and its calculation using exponential notation.

$$\frac{\boldsymbol{x}_0}{\delta_0} = \begin{bmatrix} mi\omega + \dfrac{c_f + c_r}{u} & \dfrac{ac_f - bc_r}{u} + mu \\ \dfrac{ac_f - bc_r}{u} & I_{zz}i\omega + \dfrac{a^2c_f + b^2c_r}{u} \end{bmatrix}^{-1} \begin{bmatrix} c_f \\ ac_f \end{bmatrix}$$

$$= \frac{\begin{bmatrix} I_{zz}i\omega + \dfrac{a^2c_f + b^2c_r}{u} & -\dfrac{ac_f - bc_r}{u} - mu \\ -\dfrac{ac_f - bc_r}{u} & mi\omega + \dfrac{c_f + c_r}{u} \end{bmatrix} \begin{bmatrix} c_f \\ ac_f \end{bmatrix}}{(mi\omega + \dfrac{c_f + c_r}{u})(I_{zz}i\omega + \dfrac{a^2c_f + b^2c_r}{u}) - (\dfrac{ac_f - bc_r}{u} + mu)(\dfrac{ac_f - bc_r}{u})} \tag{4.62}$$

Unfortunately, the resulting expressions do not yield simple results, and are best evaluated using a computer. Additionally, the use of exponential notation introduces complex numbers to the calculation, which results in complex values in the vector $\boldsymbol{x}_0$. The magnitudes of these complex numbers can be used to find β_0/δ_0 and r_0/δ_0.

The yaw rate response of the understeering vehicle is shown in Figure 4.12. Note that in the understeer configuration, there is a yaw natural frequency

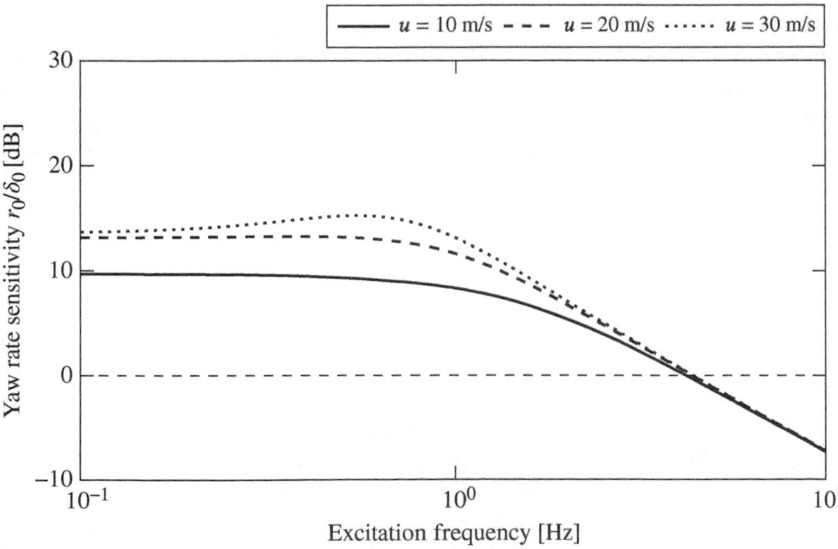

Figure 4.12 The frequency response of the yaw plane model in an understeer configuration shows a slight resonance effect around the natural frequency, but only when the speed is high enough to sufficiently reduce the yaw damping.

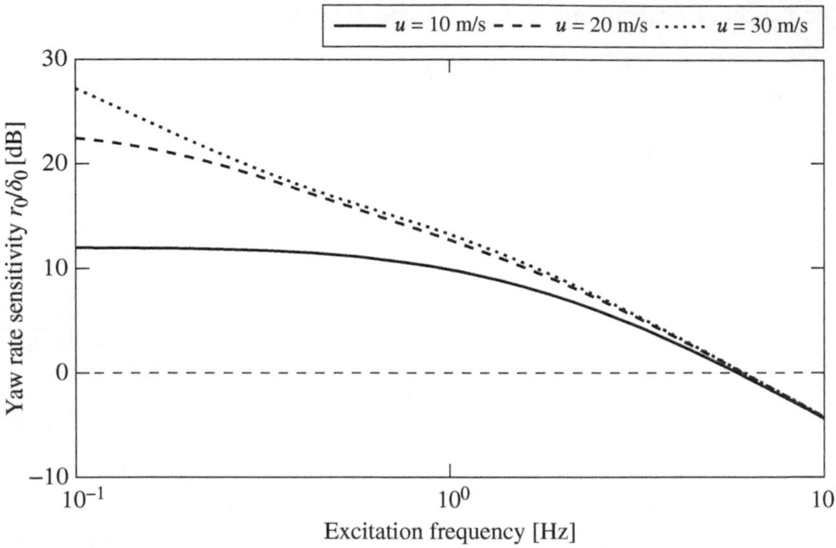

Figure 4.13 The frequency response of the yaw plane model in an oversteer configuration does not show any resonance, and has higher low frequency sensitivity.

around $\omega \approx 1$ Hz that is relatively unaffected by speed, and that the damping ratio decreases with speed. When the speed is high enough, there is a slight resonant peak around the natural frequency. The response at low frequency mimics the steady state response, i.e., initially there is an increase in sensitivity with increasing speed, reaching a maximum at the characteristic speed. Unlike the understeer configuration, the oversteer configuration, shown in Figure 4.13, has no resonance, as it is never an underdamped system. However, the rapid increase in steady state response with speed is clear when examining the low frequency behaviour.

4.1.4 Time History Solution

It is possible using the numerical methods described in Appendix A to find a time history solution to the equations of motion. This allows one to plot the path the vehicle follows in response to a particular steering input. To do so, the differential relationships between the position coordinates and the velocities must be included with the equations of motion. Using a ground fixed coordinate system, with the z axis positive in the vertical direction, and defining the *yaw angle* ψ as the angle between the vehicle and ground fixed x axes, the vehicle velocity vector can be transformed into the ground fixed axis system.

$$\dot{x} = u\cos\psi + v\sin\psi \tag{4.63}$$

$$\dot{y} = u \sin \psi - v \cos \psi \tag{4.64}$$

$$\dot{\psi} = -r \tag{4.65}$$

If the heading angle is limited to small values, the longitudinal equation reduces to simply $x = ut$. The lateral equation also simplifies, and may be combined with the heading equation and the equations of motion into a linear system. If a time history of the steer angle is provided, the system can be solved quite easily using a digital computer.

$$\begin{Bmatrix} \dot{y} \\ \dot{\psi} \\ \dot{v} \\ \dot{r} \end{Bmatrix} = \begin{bmatrix} 0 & u & -1 & 0 \\ 0 & 0 & 0 & -1 \\ 0 & 0 & -(c_f + c_r)/mu & -(ac_f - bc_r)/mu - u \\ 0 & 0 & -(ac_f - bc_r)/I_{zz}u & -(a^2 c_f + b^2 c_r)/I_{zz}u \end{bmatrix} \begin{Bmatrix} y \\ \psi \\ v \\ r \end{Bmatrix} + \begin{bmatrix} 0 \\ 0 \\ c_f/m \\ ac_f/I_{zz} \end{bmatrix} \{\delta_f\} \tag{4.66}$$

One of the somewhat surprising results that can occur in a time history solution is the appearance of a *non-minimum phase* response, particularly in the body slip angle. In response to a step or similar sudden input, a non-minimum phase system will produce a short transient that is initially in the opposite direction of the final steady state response. This occurs when the system has two real eigenvalues with somewhat differing time constants, and when the steady state response of the fast eigenvalue is smaller in magnitude and opposite in direction to the response of the slow eigenvalue. Physically, when the lateral force is applied at the front axle due to steer input, the initial lateral velocity will be in the direction of the steer input. Once the yaw motion develops, the vehicle will rotate relative to the path, and the slip angle will change direction. In the bicycle model, the ratio of the moment of inertia to the wheelbase is an indicator of how much of this behaviour one might expect. If the moment of inertia is large relative to the wheelbase, the yaw response will be slower, increasing the likelihood of the initially reverse transient. This can be expressed mathematically by checking the ratio:

$$\frac{k_{zz}^2}{ab} <> 1 \tag{4.67}$$

where k_{zz}, the yaw *radius of gyration* is defined using:

$$mk_{zz}^2 = I_{zz} \tag{4.68}$$

4.2 The Truck and Trailer Model

In this section, the equations of motion of a truck towing a trailer will be developed. A very similar approach to that of the yaw plane model will be used, but there are some notable changes, both in the model itself, and in the result. First, there are now two rigid bodies, one each for the truck and trailer, and so constraint equations must be introduced to tie the model together. More significantly, the model may now predict potentially dangerous unstable oscillatory motions that are not present in the yaw plane model. This is particularly troubling, as there continue to be numerous road accidents as a result of poorly loaded trailers, even though cause of the instability is well understood.

The first step in the development is the Newton–Euler equations of motion. The linearized equations are written in terms of the lateral speed of the truck, v, the yaw rate of the truck, r, and the sway angle of the trailer, γ. The equations of motion of the truck are written first, followed by those of the trailer. The forces acting on the truck are the same as in the yaw plane model, with the additional internal forces acting at the trailer hitch, as shown in Figure 4.14. Again, the longitudinal acceleration equation is ignored, as the truck is assumed to be moving at a constant forward speed. The trailer hitch is assumed to lie on the truck centreline, and so only the lateral force generates a moment around the centre of mass.

$$\begin{bmatrix} 1 & 1 & -1 \\ a & -b & d \end{bmatrix} \begin{Bmatrix} Y_\mathrm{f} \\ Y_\mathrm{r} \\ Y_\mathrm{h} \end{Bmatrix} = \begin{Bmatrix} m(\dot{v} + ru) \\ I_{zz}\dot{r} \end{Bmatrix} \tag{4.69}$$

The equations of motion of the trailer are written in a reference frame that is fixed to the trailer, and therefore misaligned with the frame of the truck, in which the hitch forces are defined. As a result, components of the hitch forces must be used in the force and moment expressions for the trailer. The velocities of the trailer are expressed using the same notation as for the truck, but

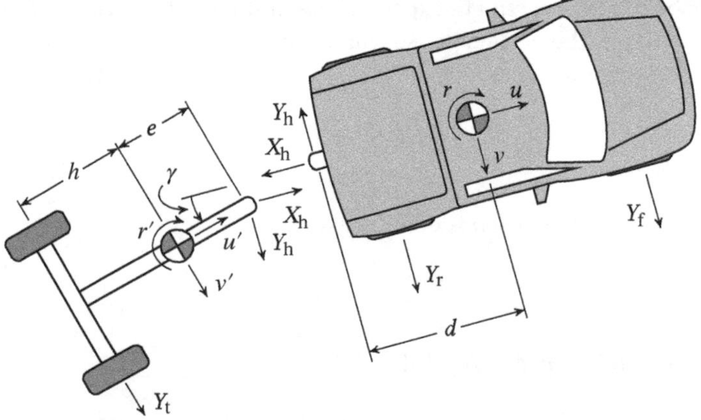

Figure 4.14 The truck and trailer model is an extension of the yaw plane model.

distinguished by the prime.

$$\begin{bmatrix} \cos\gamma & -\sin\gamma & 0 \\ \sin\gamma & \cos\gamma & 1 \\ e\sin\gamma & e\cos\gamma & -h \end{bmatrix} \begin{Bmatrix} X_\text{h} \\ Y_\text{h} \\ Y_\text{t} \end{Bmatrix} = \begin{Bmatrix} m'(\dot{u}' - r'v') \\ m'(\dot{v}' + r'u') \\ I'_{zz}\dot{r}' \end{Bmatrix} \tag{4.70}$$

Note that the sway angle can be related to the yaw rates of the truck and the trailer with the expression:

$$\dot{\gamma} = r - r' \tag{4.71}$$

In order to couple the motion of the truck with that of the trailer, a constraint must be enforced. In this case, the constraint equations state that the velocity of the trailer hitch in each direction must be the same when computed from the truck or from the trailer. The velocity of the hitch, computed in the truck reference frame, is broken into components, and equated to the velocity in the trailer reference frame.

$$\begin{Bmatrix} u' \\ v' + er' \end{Bmatrix} = \begin{bmatrix} \cos\gamma & -\sin\gamma \\ \sin\gamma & \cos\gamma \end{bmatrix} \begin{Bmatrix} u \\ v - dr \end{Bmatrix} \tag{4.72}$$

If the products of variables are neglected and small angle approximations are used, (i.e., $v\sin\gamma \approx v\gamma \approx 0,\ dr\sin\gamma \approx dr\gamma \approx 0$), the longitudinal constraint equation collapses to show that the truck and trailer have the same forward speed, as expected.

$$u' = u \tag{4.73}$$

The lateral constraint equation also simplifies somewhat, showing that the lateral speed of the trailer can be expressed as the lateral speed of the truck, plus a component of the forward speed if the trailer is misaligned, less the terms due to yaw rate of the truck and the trailer.

$$v' = u\gamma + (v - dr) - er' \tag{4.74}$$

The differentiated velocity constraints give expressions for accelerations. The constant forward speed of the truck implies the same for the trailer.

$$\dot{u}' = \dot{u} = 0 \tag{4.75}$$

The lateral expression is more complex.

$$\begin{aligned} \dot{v}' &= \dot{u}\gamma + u\dot{\gamma} + (\dot{v} - d\dot{r}) - e\dot{r}' \\ &= u\dot{\gamma} + (\dot{v} - d\dot{r}) - e\dot{r}' \end{aligned} \tag{4.76}$$

By substituting the constraint equations, using the small angle approximations, and neglecting products of variables, Equations (4.69) – (4.71) are rewritten and combined. The x component of the trailer equation only serves to show that the longitudinal hitch force X_h is negligible when treating the forward speed as constant, and is discarded. The lateral hitch force Y_{h} is eliminated through substitution.

$$\begin{bmatrix} 1 & 1 & 1 \\ a & -b & -d \\ 0 & 0 & -(e+h) \end{bmatrix} \begin{Bmatrix} Y_{\mathrm{f}} \\ Y_{\mathrm{r}} \\ Y_{\mathrm{t}} \end{Bmatrix} = \begin{Bmatrix} m(\dot{v}+ur)+m'(\dot{v}+ur-d\dot{r}-e\dot{r}') \\ (I_{zz}+m'd^2)\dot{r}-m'd(\dot{v}+ur-e\dot{r}') \\ (I'_{zz}+m'e^2)\dot{r}'-m'e(\dot{v}+ur-d\dot{r}) \end{Bmatrix} \tag{4.77}$$

The tire forces are modelled using a linear force slip relationship, where the lateral force depends only on the associated tire slip angle, in the same manner as the bicycle model.

$$\begin{Bmatrix} Y_{\mathrm{f}} \\ Y_{\mathrm{r}} \\ Y_{\mathrm{t}} \end{Bmatrix} = -\begin{bmatrix} c_{\mathrm{f}} & 0 & 0 \\ 0 & c_{\mathrm{r}} & 0 \\ 0 & 0 & c_{\mathrm{t}} \end{bmatrix} \begin{Bmatrix} \alpha_{\mathrm{f}} \\ \alpha_{\mathrm{r}} \\ \alpha_{\mathrm{t}} \end{Bmatrix} \tag{4.78}$$

Kinematic expressions are used to find the slip angles.

$$\begin{aligned} \alpha_{\mathrm{t}} &= \frac{v'-hr'}{u'} \\ &= \frac{u\gamma+(v-dr)-(e+h)r'}{u} \\ &= \gamma + \frac{(v-dr)-(e+h)r'}{u} \end{aligned} \tag{4.79}$$

$$\begin{Bmatrix} \alpha_{\mathrm{f}} \\ \alpha_{\mathrm{r}} \\ \alpha_{\mathrm{t}} \end{Bmatrix} = \frac{1}{u}\begin{bmatrix} 0 & 1 & a & 0 \\ 0 & 1 & -b & 0 \\ u & 1 & -d & -(e+h) \end{bmatrix} \begin{Bmatrix} \gamma \\ v \\ r \\ r' \end{Bmatrix} - \begin{Bmatrix} \delta_{\mathrm{f}} \\ 0 \\ 0 \end{Bmatrix} \tag{4.80}$$

The equations are coupled and first order in the truck lateral speed and yaw rate, and trailer yaw rate, but also include sway angle terms. In order to produce a complete first order set, they are complemented with Equation (4.71). The equations of motion become:

$$\begin{bmatrix} 1 & 0 & 0 & 0 \\ 0 & m+m' & -m'd & -m'e \\ 0 & -m'd & I_{zz}+m'd^2 & m'ed \\ 0 & -m'e & m'ed & I'_{zz}+m'e^2 \end{bmatrix} \begin{Bmatrix} \dot{\gamma} \\ \dot{v} \\ \dot{r} \\ \dot{r}' \end{Bmatrix} +$$

$$\frac{1}{u}\begin{bmatrix} 0 & 0 & -u & u \\ uc_{\mathrm{t}} & c_{\mathrm{f}}+c_{\mathrm{r}}+c_{\mathrm{t}} & (m+m')u^2+ac_{\mathrm{f}}-bc_{\mathrm{r}}-dc_{\mathrm{t}} & -(e+h)c_{\mathrm{t}} \\ -duc_{\mathrm{t}} & ac_{\mathrm{f}}-bc_{\mathrm{r}}-dc_{\mathrm{t}} & -m'du^2+a^2c_{\mathrm{f}}+b^2c_{\mathrm{r}}+d^2c_{\mathrm{t}} & (e+h)dc_{\mathrm{t}} \\ -(e+h)uc_{\mathrm{t}} & -(e+h)c_{\mathrm{t}} & -m'eu^2+(e+h)dc_{\mathrm{t}} & (e+h)^2c_{\mathrm{t}} \end{bmatrix} \begin{Bmatrix} \gamma \\ v \\ r \\ r' \end{Bmatrix}$$

$$= \begin{bmatrix} 0 \\ c_{\mathrm{f}} \\ ac_{\mathrm{f}} \\ 0 \end{bmatrix} \{\delta_{\mathrm{f}}\} \tag{4.81}$$

and are now in the form:

$$\mathbf{M}\dot{\boldsymbol{x}} + \mathbf{L}\boldsymbol{x} = \mathbf{F}\ \delta_{\mathrm{f}} \tag{4.82}$$

This is exactly as in the yaw plane model, and accordingly, a similar set of analyses can be conducted.

4.2.1 Steady State Analysis

The form of the truck and trailer model allows the same type of steady state analysis.

$$\boldsymbol{x}/\delta_{\mathrm{f}} = \mathbf{L}^{-1}\mathbf{F} \tag{4.83}$$

Computation of the inverse of the matrix $\mathbf{L}$ is lengthy, but allows a closed form solution. Note that imposing a steady state solution on the sway angle forces the yaw rate of the truck and trailer to be the same.

$$\frac{r}{\delta_{\mathrm{f}}} = \frac{r'}{\delta_{\mathrm{f}}} = \frac{u}{a + b - \dfrac{u^2}{(a+b)c_{\mathrm{f}}c_{\mathrm{r}}}\left(m(ac_{\mathrm{f}} - bc_{\mathrm{r}}) + m'\dfrac{h}{e+h}((a+d)c_{\mathrm{f}} + (d-b)c_{\mathrm{r}})\right)} \tag{4.84}$$

As in the yaw plane vehicle case, the yaw rate gain determines oversteer or understeer. In order to have an understeering truck and trailer combination, then the denominator must never tend to zero, or:

$$m(ac_{\mathrm{f}} - bc_{\mathrm{r}}) + m'\frac{h}{e+h}((a+d)c_{\mathrm{f}} + (d-b)c_{\mathrm{r}}) < 0 \tag{4.85}$$

An examination of this expression shows that if an understeering truck is fitted with a trailer with $h = 0$, i.e., the trailer axle lies under its mass centre, then the truck and trailer combination will also understeer. However, as will be seen, this configuration can produce other unexpected undesirable results. Rewriting Equation (4.85) gives:

$$(m(e+h) + m'h)(ac_{\mathrm{f}} - bc_{\mathrm{r}}) + m'dh(c_{\mathrm{f}} + c_{\mathrm{r}}) < 0 \tag{4.86}$$

i.e., if the trailer's centre of mass is in front of its axle ($h > 0$), but the trailer hitch is located at the truck's mass centre ($d = 0$), then an understeering truck would again continue to understeer when fitted with the trailer. Of course, this configuration is less practical. If Equation (4.85) does not hold, then the critical speed can be found from:

$$a+b-\frac{u^2}{(a+b)c_f c_r}\left(m(ac_f-bc_r)+m'\frac{h}{e+h}((a+d)c_f+(d-b)c_r)\right)=0 \tag{4.87}$$

or:

$$u_{\text{crit}}=\sqrt{\frac{(a+b)^2c_f c_r}{m(ac_f-bc_r)+m'\frac{h}{e+h}((a+d)c_f+(d-b)c_r)}} \tag{4.88}$$

Expressions for the steady state body slip angle and steady state trailer sway angle gains can also be found, but both are too elaborate to be particularly illustrative. It is noteworthy that if the trailer mass is set to zero, the expression for the body slip angle gain reduces to match exactly the expression given by the yaw plane model (Equation (4.33)).

$$\frac{\beta}{\delta_f}=\frac{b-\frac{u^2}{(a+b)c_r}\left(ma+m'\frac{h}{e+h}(a+d)\right)}{a+b-\frac{u^2}{(a+b)c_f c_r}\left(m(ac_f-bc_r)+m'\frac{h}{e+h}((a+d)c_f+(d-b)c_r)\right)} \tag{4.89}$$

$$\frac{\gamma}{\delta_f}=\frac{d-b+e+h+u^2\left(\frac{m}{c_r}\frac{a}{a+b}-\frac{m'}{c_t}\frac{e}{e+h}+\frac{m'}{c_r}\frac{h(a+d)}{(a+b)(e+h)}\right)}{a+b-\frac{u^2}{(a+b)c_f c_r}\left(m(ac_f-bc_r)+m'\frac{h}{e+h}((a+d)c_f+(d-b)c_r)\right)} \tag{4.90}$$

The values of the steady state gains are plotted using the parameters in Table 4.2 and shown in Figure 4.15. The parameters used in the study are chosen to represent the same vehicle used in the yaw plane model, only with the centre of mass moved slightly back (0.1 m) to promote a less understeering behaviour, combined with a relatively large trailer. In Figure 4.16, the plots are repeated with the centre of mass of the trailer shifted forward 0.4 m, keeping the total trailer length $e+h$ constant, to demonstrate the effect of the critical speed.

4.2.2 Transient Analysis

As in the yaw plane model, the stability of the truck and trailer can be evaluated by performing an eigen analysis. The equations of motion can be cast into standard form by multiplying by the inverse of the mass matrix; then the eigenvalues of $-\mathbf{M}^{-1}\mathbf{L}$ determine the stability of the system. With a 4×4 matrix, the eigenvalues are extremely challenging to find by hand, but if one resorts to using a numerical method on a digital computer, they can be found easily. Two

Table 4.2 Truck and trailer parameter values

Quantity	Notation	Value
truck front axle dimension	a	1.289 m
truck rear axle dimension	b	1.596 m
truck hitch dimension	d	2.7 m
trailer hitch dimension	e	2.9 m
trailer axle dimension	h	0.1 m
truck mass	m	1 730 kg
truck yaw inertia	I_{zz}	3 508 kgm^2
trailer mass	m'	2 000 kg
trailer yaw inertia	I'_{zz}	3 000 kgm^2
front tire cornering stiffness	c_f	80 000 N m/rad
rear tire cornering stiffness	c_r	80 000 N m/rad
trailer tire cornering stiffness	c_t	80 000 N m/rad

Note: the value of the cornering stiffness is a sum of the left and right side for each axle.

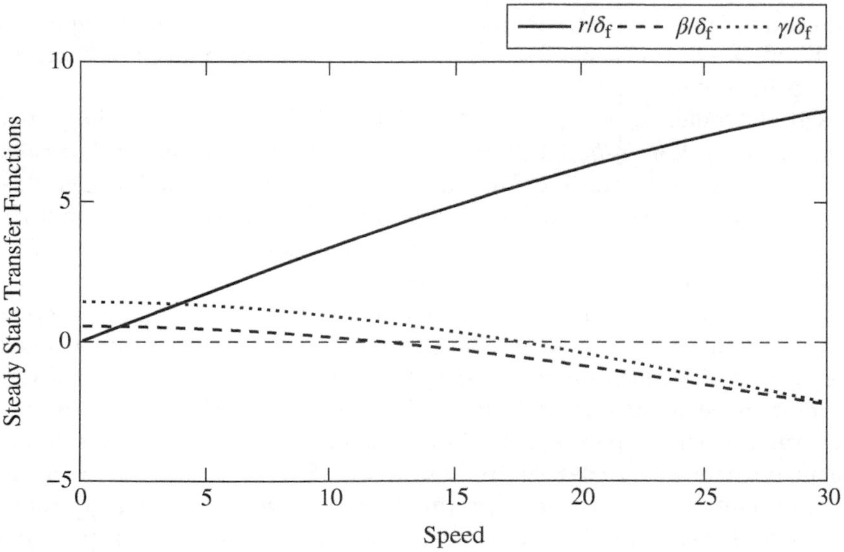

Figure 4.15 The steady state gains of the truck-trailer model in the default configuration do not give any indication of instability.

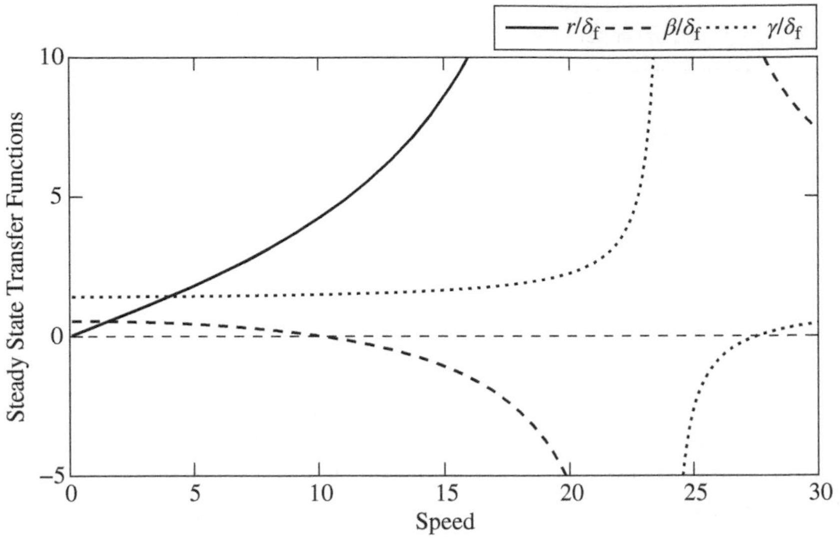

Figure 4.16 The steady state gains of the truck-trailer model with the trailer's centre of mass shifted forward display an asymptote at the critical speed, much like the oversteer condition in the yaw plane model.

configurations are analyzed here, with the only difference being a forward shift in the location of the centre of mass of the trailer. As it happens, the location of the centre of mass of the trailer relative to its axle proves to be very important in the truck and trailer model. The vehicle parameter data given in Table 4.2 was again used for the default analysis.

With the trailer's centre of mass close to its axle (0.1 m in front), the system has an unstable *fish-tailing* oscillatory mode, indicated by a complex eigenvalue with a positive real part. The damping in the oscillatory mode has a gradual decrease until the system becomes unstable at a forward speed of approximately 20 m/s. As the name implies, the fish-tailing mode is characterized by large amounts of trailer sway, in this case at $\omega \approx 0.6$ Hz.

With the centre of mass of the trailer shifted forward 0.4 m, but keeping the total trailer length $e + h$ constant, and all other parameters unchanged, the behaviour is much different. Although the oscillatory mode is still very lightly damped, it remains stable. Instead, the system now shows a positive real eigenvalue when the speed reaches 24 m/s, indicating a *jack-knifing* instability. Again, the name is very descriptive. The jack-knifing mode is characterized by large amounts of yaw in the truck; in this mode, the two bodies 'fold up' around the trailer hinge. Note that in both these unstable configurations, the truck alone would have been stable. The addition of the trailer causes the instability. Note also that a region of stability exists between these two configurations. In general, increasing the forward speed, or the mass of the trailer, shrinks the region of stability for the location of the trailer's centre of mass.

The eigenvalue calculation results are presented in Figures 4.17 and 4.18. When comparing these to the steady state transfer functions shown in

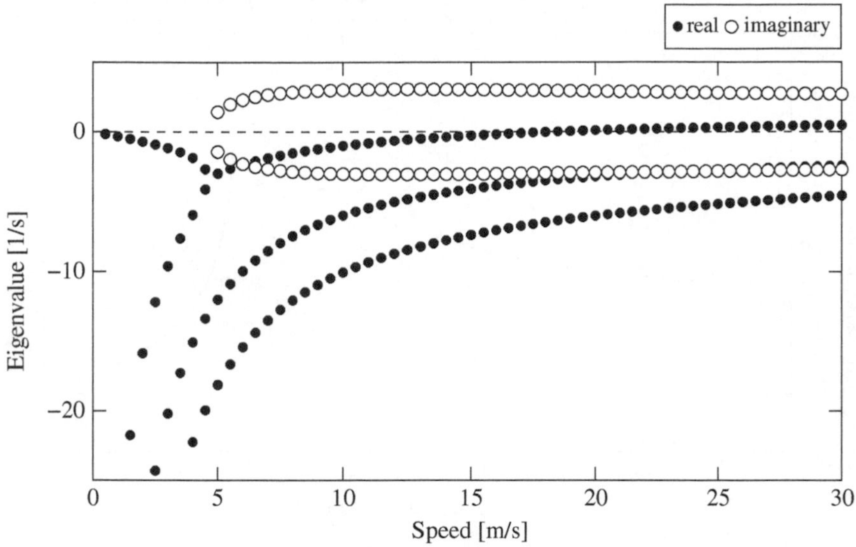

Figure 4.17 In a truck and trailer, when the centre of mass of the trailer is too far rearward, a fish-tailing mode can develop. This motion is characterized by large trailer oscillations of approximately constant frequency. The damping in the sway decreases with speed, and becomes unstable above some critical speed.

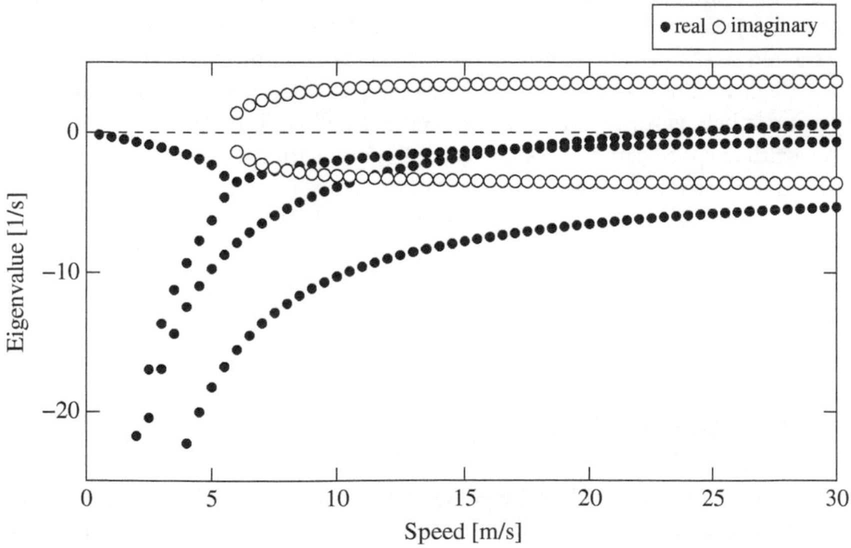

Figure 4.18 In the same truck and trailer, if the trailer's centre of mass is pushed forward, the oscillatory motion will be more damped and remain stable, but another instability called jack-knifing can appear. The jack-knifing mode is not oscillatory, but rather exponential growth, consisting of mainly truck yaw motion, and is also unstable above some critical speed.

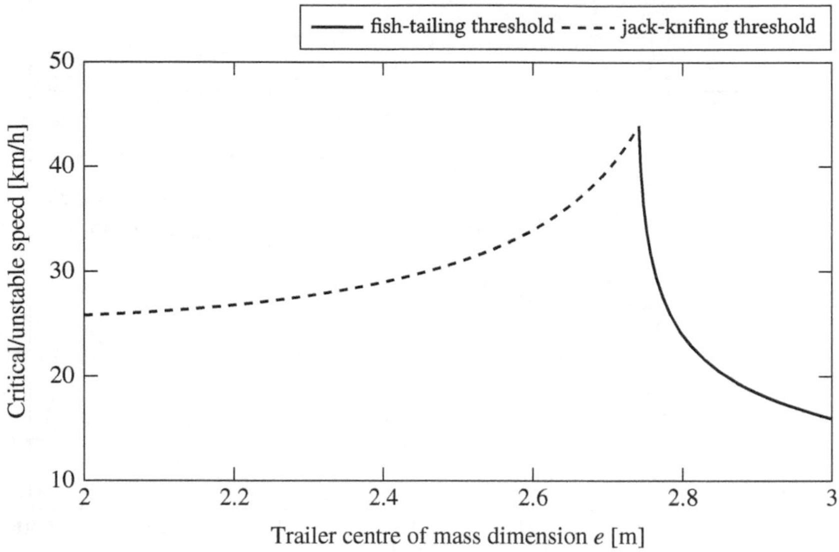

Figure 4.19 The truck-trailer critical speed can be found for a range of values of the trailer centre of mass location. The size of the stable region decreases as forward speed increases. Note that in this case, the total trailer length from the hitch to the axle was held constant at $e + h = 3$ m, i.e., as e was increased, h was decreased by the same amount.

Figures 4.15 and 4.16, an important feature is noted. The fish-tailing instability cannot be predicted through the study of the steady state behaviour and is not related to the critical speed defined there. For the nonoscillatory unstable mode (jack-knifing), the steady state transfer functions show a very distinct asymptotic behaviour in the region of the instability. The closer the vehicle is to the critical speed, the more sensitive it becomes to steering inputs, in effect, warning the driver. However, if the instability is oscillatory, no such behaviour exists, and the change in sensitivity is very gradual. This may have an influence in driver behaviour, reducing the ability to detect the onset of oscillatory instability. Figure 4.19 shows the range of stable values for the location of the centre of mass of the trailer.

4.3 The Quarter Car Model

The quarter car model has been used for many years for predicting ride quality. It is a simple two degree of freedom model, with two bodies constrained to vertical translation, representing the sprung mass m_s (the chassis, powertrain, driver, cargo, etc.) and the unsprung mass m_u (the wheel, hub, brake rotor or drum, etc.). A schematic diagram is shown in Figure 4.20.

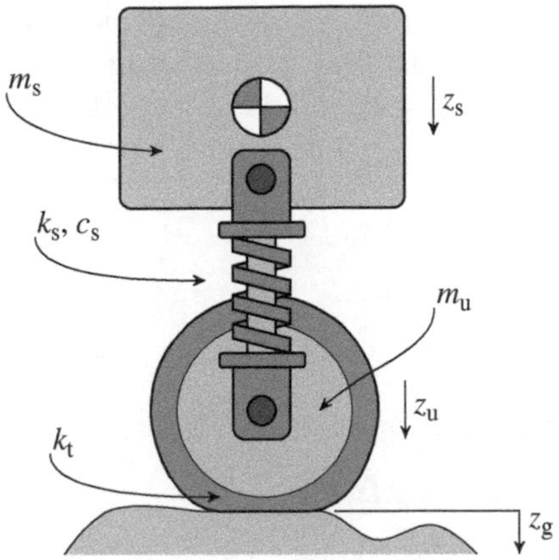

Figure 4.20 The quarter car model is a simplification that classifies the vehicle components as either sprung or unsprung mass.

The bodies are attached by a linear spring and damper representing the suspension, with coefficients k_s and c_s, and the wheel is held to the ground by a spring that represents the tire elasticity, with the stiffness denoted as k_t. Often, the damping in the tire is ignored, as in practice it tends to be very small, and its inclusion adds significantly to the complexity of the resulting model. The displacements of the sprung mass and unsprung mass are written as z_s and z_u respectively, while z_g represents the ground height, all measured from a fixed inertial reference. The model is formed by writing Newton's equation of motion for each of the two bodies. For the sprung mass:

$$m_s\ddot{z}_s + c_s(\dot{z}_s - \dot{z}_u) + k_s(z_s - z_u) = 0 \tag{4.91}$$

For the unsprung mass:

$$m_u\ddot{z}_u + c_s(\dot{z}_u - \dot{z}_s) + k_s(z_u - z_s) + k_t(z_u - z_g) = 0 \tag{4.92}$$

The two equations are combined as a vector equation:

$$\begin{bmatrix} m_s & 0 \\ 0 & m_u \end{bmatrix} \begin{Bmatrix} \ddot{z}_s \\ \ddot{z}_u \end{Bmatrix} + \begin{bmatrix} c_s & -c_s \\ -c_s & c_s \end{bmatrix} \begin{Bmatrix} \dot{z}_s \\ \dot{z}_u \end{Bmatrix} + \begin{bmatrix} k_s & -k_s \\ -k_s & k_s + k_t \end{bmatrix} \begin{Bmatrix} z_s \\ z_u \end{Bmatrix} = \begin{bmatrix} 0 \\ k_t \end{bmatrix} \{z_g\} \tag{4.93}$$

or:

$$\mathbf{M}\ddot{\mathbf{z}} + \mathbf{L}\dot{\mathbf{z}} + \mathbf{K}\mathbf{z} = \mathbf{F}\boldsymbol{u} \tag{4.94}$$

4.3.1 Transient Analysis

In the transient condition, after a vehicle has been disturbed, its vertical motion is mostly oscillatory. The quarter car model is often used to determine the frequencies of motion of both the vehicle body, and the unsprung mass. When the equations are manipulated by hand, the damping terms are often omitted, as they have a relatively small effect on the result, and this simplification allows a sinusoidal solution to the equations. Suppose that the road is smooth (i.e., $\boldsymbol{u} = \{z_g\} = 0$), and that the motions can be written:

$$\boldsymbol{z} = \begin{Bmatrix} z_s \\ z_u \end{Bmatrix} = \boldsymbol{z}_0 \cos \omega t \tag{4.95}$$

and as a result:

$$\ddot{\boldsymbol{z}} = -\omega^2 \boldsymbol{z}_0 \cos \omega t = -\omega^2 \boldsymbol{z} \tag{4.96}$$

The equation of motion is then:

$$-\omega^2 \mathbf{M}\boldsymbol{z} + \mathbf{K}\boldsymbol{z} = \mathbf{0} \tag{4.97}$$

or:

$$[\mathbf{K} - \omega^2 \mathbf{M}]\boldsymbol{z} = \mathbf{0} \tag{4.98}$$

With an examination of the resulting equation, one should recognize that it is effectively an eigenvector problem, and therefore must have a singular coefficient matrix. Expanding the determinant of this matrix term gives the characteristic equation:

$$(k_s - \omega^2 m_s)(k_s + k_t - \omega^2 m_u) - k_s^2 = 0 \tag{4.99}$$

The characteristic equation appears to be quartic, but a substitution ($\omega^2 = s$) reveals that it is really only quadratic, as expected in a system of two equations.

$$m_s m_u s^2 - ((k_s + k_t)m_s + k_s m_u)s + k_s k_t = 0 \tag{4.100}$$

At this point, a solution from the quadratic would be relatively straightforward, but the resulting expressions are not particularly compact or convenient. Instead, some simplifications are made. When the typical values of the mass and stiffness are used, the sprung mass will be roughly an order of magnitude more than the unsprung mass, and similarly, the tire stiffness is roughly an order of magnitude more than the suspension stiffness. As a result, $k_t m_s \gg k_s m_u$. This allows the linear coefficient of the quadratic to be safely modified.

$$(k_s + k_t)m_s + k_s m_u \approx (k_s + k_t)m_s \tag{4.101}$$

Finally, the first and last terms in the quadratic can be alternately ignored (in the case that s is small, or large, respectively), allowing two simple linear equations that can be solved for s, and in turn for ω. First:

$$m_s m_u s - (k_s + k_t)m_s = 0 \tag{4.102}$$

$$\omega = \sqrt{\frac{k_s + k_t}{m_u}} \tag{4.103}$$

and second:

$$-(k_s + k_t)m_s s + k_s k_t = 0 \tag{4.104}$$

$$\omega = \sqrt{\frac{k_s k_t}{(k_s + k_t)m_s}} \tag{4.105}$$

An examination of the two results shows two frequencies of motion; the lower is typically on the order of about 2π rad/s or 1 Hz, while the higher is around 20π rad/s or 10 Hz. The resulting expression for the higher frequency shows that it is equivalent to a system where the sprung mass is held fixed while the unsprung mass bounces against the suspension and tire as two parallel springs. The low frequency expression is equivalent to a system where the unsprung mass is ignored, and the sprung mass bounces against the suspension and tire as two springs in series. The simplified models are illustrated in Figure 4.21. The two motions are typically labelled the *wheel hop* mode and the *bounce* or *heave* mode, respectively.

One of the common confusions that arises from the quarter car model is the notion that each frequency in a multi-degree of freedom system is associated with the motion of a particular body. In general, this is not the case at all. All the bodies in the system will vibrate at each of the natural frequencies, but some will move more than others at each frequency. It is just a property of the quarter car model that the motions tend to be very discrete; the low frequency is associated with the vehicle body and the high frequency with the wheel. However, in a general vibrating mechanical system, this one-to-one attachment of frequencies to bodies does not exist.

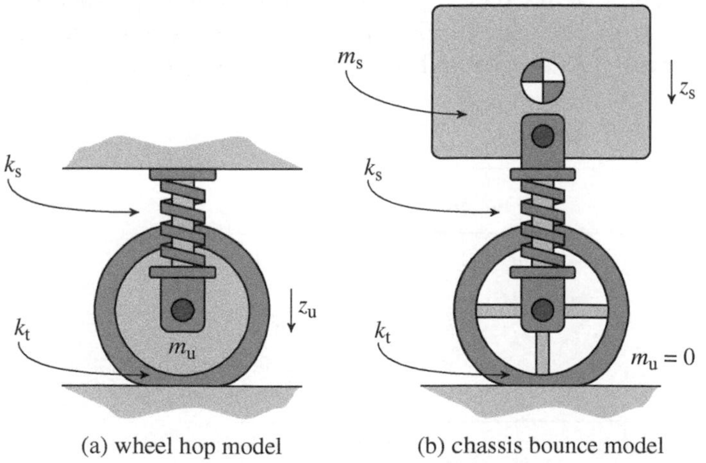
k_s
k_t
m_u
z_u
m_s
z_s
k_s
k_t
$m_u = 0$
(a) wheel hop model
(b) chassis bounce model

Figure 4.21 Due to the two widely separate natural frequencies, the quarter car model can be replaced by two single degree of freedom models with only a small loss in fidelity.

Example

Using the values in Table 4.3, the exact solution of the natural frequencies can be compared to the simplified solution, using Equations (4.102) – (4.105).

$$\omega_1 = \sqrt{\frac{k_s + k_t}{m_u}} = \sqrt{\frac{198000}{50}} = 62.9 \text{ rad/s} = 10.0 \text{ Hz}$$

$$\omega_2 = \sqrt{\frac{(180000)(18000)}{(198000)(500)}} = 5.72 \text{ rad/s} = 0.910 \text{ Hz}$$

To find the exact solutions, the matrix characteristic equation is solved.

$$\det[\mathbf{M}\,\omega^2 - \mathbf{K}] = \det\begin{bmatrix} 500\omega^2 - 18000 & 18000 \\ 18000 & 50\omega^2 - 198000 \end{bmatrix} = 0$$

$$(500\omega^2 - 18000)(50\omega^2 - 198000) - 18000^2 = 0$$

$$\begin{aligned} \omega &= 63.0 \text{ rad/s}, 5.72 \text{ rad/s} \\ &= 10.0 \text{ Hz}, 0.910 \text{ Hz} \end{aligned}$$

Clearly, the approximate solution is a very close to the exact solution. However, one advantage of the exact solution is that the associated amplitudes of the motion can also be found. Taking $\omega = 63.0$:

$$[\mathbf{M}\,\omega^2 - \mathbf{K}]\boldsymbol{x}_0 = \begin{bmatrix} 1963650 & 18000 \\ 18000 & 165 \end{bmatrix} \boldsymbol{x}_0 = 0$$

An inspection will show that the rows of the matrix are linear multiples of each other. One entry in $\boldsymbol{x}_0$ can be chosen arbitrarily, and either equation can be solved for the other entry. As a result, only the *relative* size of the motions can

Table 4.3 Quarter car parameter values

Quantity	Notation	Value
sprung mass	m_s	500 kg
unsprung mass	m_u	50 kg
suspension stiffness	k_s	18 000 N/m
suspension damping	c_s	1 000 Ns/m
tire stiffness	k_t	180 000 N/m

Note: the value of the sprung mass is a fraction of the sprung mass of the entire vehicle, typically a quarter, but sometimes adjusted slightly to account for the front/rear weight bias.

be found, not the actual size. Setting the first entry to a value of one gives a solution.

$$\boldsymbol{x}_0 = \boldsymbol{x}_1 = \begin{Bmatrix} 1 \\ -18000/165 \end{Bmatrix} = \begin{Bmatrix} 1 \\ -109 \end{Bmatrix}$$

In the high frequency mode, the motion of the sprung mass is about one hundred times as large as the unsprung mass motion, and the motions are in opposite directions. Taking $\omega = 5.72$:

$$[\mathbf{M}\ \omega^2 - \mathbf{K}]\boldsymbol{x}_0 = \begin{bmatrix} -1650 & 18000 \\ 18000 & -196365 \end{bmatrix} \boldsymbol{x}_0 = 0$$

Choosing the second entry as one gives the other solution.

$$\boldsymbol{x}_0 = \boldsymbol{x}_2 = \begin{Bmatrix} 18000/1650 \\ 1 \end{Bmatrix} = \begin{Bmatrix} 10.9 \\ 1 \end{Bmatrix}$$

In the low frequency mode, the motion of the sprung mass is about eleven times as large as the unsprung mass motion, and the motions are in the same direction. The relative magnitudes of the motions reinforce the soundness of the assumptions in the simplification. Finally, note that the exact solution is effectively the eigenvalues and eigenvectors of the matrix $\mathbf{M}^{-1}\mathbf{K}$; these can be calculated directly using a digital computer.

Effect of Damping

The transient solution of the quarter car model can be solved with relative ease when damping is ignored. If one wishes to include the effects of damping, then the problem becomes somewhat more challenging, and is not easily tractable by hand. If one turns to a digital computer to proceed, a better alternative is to reduce the equations to a first order or *state space* form, as shown in Chapter 7. In this case, there are now four eigenvalues to solve, and they will arrive as complex conjugate pairs. Both the natural frequency and damping ratio information can be found from the complex eigenvalues. Similarly, the eigenvectors will also become complex, as they will contain information not only about the relative size of the motions, but also their phasing, which is no longer restricted to either 0° (in phase) or 180° (out of phase). The form of the first order equations is:

$$\dot{\boldsymbol{x}} = \mathbf{A}\boldsymbol{x} + \mathbf{B}\boldsymbol{u} \tag{4.106}$$

In this case, the following definitions are used:

$$\boldsymbol{x} = \begin{Bmatrix} \boldsymbol{z} \\ \dot{\boldsymbol{z}} \end{Bmatrix} = \begin{Bmatrix} z_s \\ z_u \\ \dot{z}_s \\ \dot{z}_u \end{Bmatrix}, \quad \boldsymbol{u} = z_g \tag{4.107}$$

and:

$$\mathbf{A} = \begin{bmatrix} \mathbf{0} & \mathbf{I} \\ -\mathbf{M}^{-1}\mathbf{K} & -\mathbf{M}^{-1}\mathbf{L} \end{bmatrix}, \quad \mathbf{B} = \begin{bmatrix} \mathbf{0} \\ \mathbf{M}^{-1}\mathbf{F} \end{bmatrix} \tag{4.108}$$

Example

Again, using the values in Table 4.3, Equation (4.108) becomes:

$$\begin{Bmatrix} \dot{z}_s \\ \dot{z}_u \\ \ddot{z}_s \\ \ddot{z}_u \end{Bmatrix} = \begin{bmatrix} 0 & 0 & 1 & 0 \\ 0 & 0 & 0 & 1 \\ -36 & 36 & -2 & 2 \\ 360 & -3960 & 20 & -20 \end{bmatrix} \begin{Bmatrix} z_s \\ z_u \\ \dot{z}_s \\ \dot{z}_u \end{Bmatrix} + \begin{bmatrix} 0 \\ 0 \\ 0 \\ 3600 \end{bmatrix} z_g$$

The eigenvalues of the **A** matrix are:

$$s = -10.2 \pm 61.9\text{i}, -0.831 \pm 5.68\text{i}$$

The magnitudes of the eigenvalues give the natural frequencies.

$$\omega_1 = \sqrt{10.2^2 + 61.9^2} = 62.7 \text{ rad/s} = 9.98 \text{ Hz}$$

$$\omega_2 = \sqrt{0.831^2 + 5.68^2} = 5.74 \text{ rad/s} = 0.914 \text{ Hz}$$

The results are quite close to those predicted when damping is ignored. The additional information in the model also allows damping ratios to be found from the eigenvalues.

$$\zeta_1 = \frac{10.2}{62.7} = 0.162$$

$$\zeta_2 = \frac{0.831}{5.74} = 0.145$$

In this case, both modes are lightly damped, as is typical in most automotive suspensions. The eigenvectors of the **A** matrix are:

$$\boldsymbol{x}_0 = \begin{Bmatrix} -0.000427 \pm 0.000284\text{i} \\ -0.00259 \pm 0.0157\text{i} \\ -0.0132 \pm 0.0293\text{i} \\ 0.999 \end{Bmatrix}, \begin{Bmatrix} -0.0247 \pm 0.169\text{i} \\ 0.00233 \pm 0.0156\text{i} \\ 0.981 \\ 0.0868 \pm 0.0262\text{i} \end{Bmatrix}$$

Because of the way the equations are reduced to first order, the eigenvectors display an interesting property. Consider the top two rows of the first eigenvector, multiplied by the corresponding eigenvalue.

$$(-10.2 + 61.9\text{i}) \begin{Bmatrix} -0.000427 + 0.000284\text{i} \\ -0.00259 - 0.0157\text{i} \end{Bmatrix} = \begin{Bmatrix} -0.0132 - 0.0293\text{i} \\ 0.999 \end{Bmatrix}$$

The result is exactly the bottom two rows, which implies that the bottom half of each eigenvector contains the same information as the top, so one can be discarded as redundant. Scaling the bottom two entries, and rewriting in magnitude and phase form, gives:

$$\boldsymbol{x}_1 = \begin{Bmatrix} 1\angle 0^\circ \\ 31.1\angle 114^\circ \end{Bmatrix}$$

In the high frequency mode, the unsprung mass motion is significantly larger, and largely out of phase. For the low frequency mode:

$$\boldsymbol{x}_2 = \begin{Bmatrix} 10.8\angle -16.8^\circ \\ 1\angle 0^\circ \end{Bmatrix}$$

The sprung mass motion is much larger, and nearly in phase. A comparison of the relative size of the motions for each of the two vectors to the undamped case shows a relatively small change.

4.3.2 Frequency Response

Once the natural frequencies of motion have been determined, and one has a sense of the type of motions that occur in a suspension, further explorations can be done by considering the quarter car model as a forced motion system, and finding the frequency response. In this case the input is usually the road disturbance, and the output may be either the sprung mass motion, or perhaps the relative motion between the sprung and unsprung mass, as shown in Figure 4.22. If the objective of the study is improved ride comfort, one might choose the sprung mass acceleration as the output, as it is quite indicative of

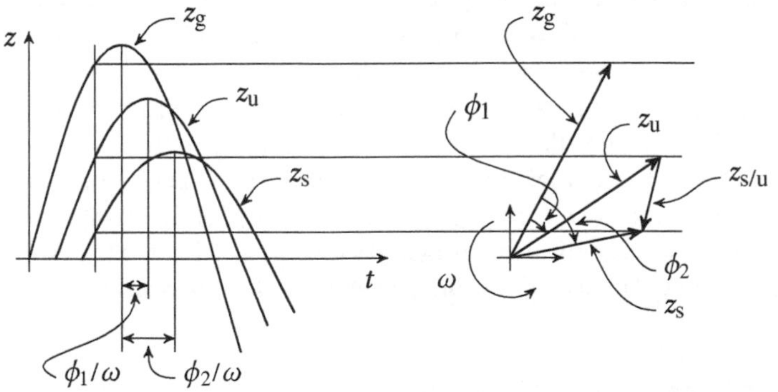

Figure 4.22 The sinusoidal motion of the quarter car model is conceptually identical to three vectors rotating with angular speed ω, with the relative lengths and angular phasing changing as a function of the angular speed. In this case, the relative displacement between the sprung and unsprung mass $z_{s/u}$ represents the suspension travel.

passenger discomfort. Choosing the suspension parameters to reduce acceleration can give improvements in ride quality. Alternately, the relative motion between the wheel and the ground, i.e., the tire compression, may be chosen as the model output, as it is a good indicator of tire grip. Minimizing the amplitude of the tire force oscillations results in a reduced likelihood that the tire may lose traction on the road, and thus improvements in lateral and longitudinal performance are gained.

In the forced motion analysis, the road is assumed to have a sinusoidal profile. While it may initially seem odd to consider the road as sinusoidal, this follows from the concept of *Fourier analysis*, which states that any repeating waveform can be composed of a linear combination of sine waves, over an infinite span of frequencies. The road roughness will be modelled as though it is composed of a sum of these sinusoids at different amplitudes and wavelengths. The resulting motion of the vehicle can be found by summing the response to each component, i.e., a linear superposition. In practice, the wavelengths considered might range from as short as 0.1 m to as long as 100 m. The international standard ISO 8606 defines a relationship between the amplitudes and the wavelengths found in various classes of roads using a technique called the *power spectral density*. Using ISO 8606, a random road profile can be constructed using a sum similar to:

$$z_g(x) = \sum_{i=1}^{n} \frac{z_0}{i} \cos\left(\frac{2\pi i}{\lambda_0} x + \phi_i\right) \tag{4.109}$$

where $z_i = z_0/i$ is the amplitude of each component, $\lambda_i = \lambda_0/i$ is the wavelength of each component, ϕ_i is a random phase angle between 0 and 2π, and n is the number of terms in the series. The base wavelength λ_0 is generally equal to the length of road being modelled, and the base amplitude z_0 depends on both the base wavelength and the class of road; for a 100 m road of mediocre quality, a value of $z_0 = 0.032$ m would be reasonable. For details, see Rill [1].

To proceed with the frequency response analysis, the damping must be considered in the model. While the effect of ignoring the damping on the frequency of the transient motion is quite small, the same cannot be said for the amplitude of forced motion, where it can be very significant. As a result of the damping in the system, neither the sprung nor unsprung mass motion will be in phase with the disturbance, so an exponential notation is preferred. The road amplitude is written as:

$$z_g(t) = z_{g0} e^{i\omega t} = z_{g0}(\cos\omega t + i \sin\omega t) \tag{4.110}$$

and the output is:

$$z(t) = z_0 e^{i\omega t} \tag{4.111}$$

In order to complete the frequency response calculation, one of two approaches is necessary. One option is to continue the analysis in the second order form.

$$-\omega^2 \mathbf{M} \boldsymbol{z}_0 + \mathrm{i}\omega \mathbf{L} \boldsymbol{z}_0 + \mathbf{K} \boldsymbol{z}_0 = \mathbf{F} \boldsymbol{z}_{g0} \tag{4.112}$$

$$\boldsymbol{z}_0 = [-\omega^2 \mathbf{M} + \mathrm{i}\omega \mathbf{L} + \mathbf{K}]^{-1} \mathbf{F} \boldsymbol{z}_{g0} \tag{4.113}$$

$$z_0/z_{g0} = \begin{bmatrix} -\omega^2 m_s + \mathrm{i}\omega c_s + k_s & -\mathrm{i}\omega c_s - k_s \\ -\mathrm{i}\omega c_s - k_s & -\omega^2 m_u + \mathrm{i}\omega c_s + k_s + k_t \end{bmatrix}^{-1} \begin{bmatrix} 0 \\ k_t \end{bmatrix} \{z_g\} \tag{4.114}$$

Unfortunately, expanding the inverse of the resulting 2×2 matrix does not offer any significant simplification. The system is at least in a form where hand computation is possible, but the presence of complex numbers complicates matters. Again, a reduction to first order form, and the use of a computer offers an advantage. In this case, the first order equations take the form:

$$\begin{Bmatrix} \dot{\boldsymbol{x}} \\ \boldsymbol{y} \end{Bmatrix} = \begin{bmatrix} \mathbf{A} & \mathbf{B} \\ \mathbf{C} & \mathbf{D} \end{bmatrix} \begin{Bmatrix} \boldsymbol{x} \\ \boldsymbol{u} \end{Bmatrix} \tag{4.115}$$

where the **A** and **B** matrices are defined as in Equation (4.108). The addition of the output equation $\boldsymbol{y} = \mathbf{C}\boldsymbol{x} + \mathbf{D}\boldsymbol{u}$ allows some interesting additional information to be obtained. The output matrices are:

$$\mathbf{C} = \begin{bmatrix} 1 & 0 & 0 & 0 \\ 1 & -1 & 0 & 0 \\ 0 & 1 & 0 & 0 \end{bmatrix}, \ \mathbf{D} = \begin{bmatrix} 0 \\ 0 \\ -1 \end{bmatrix} \tag{4.116}$$

Choosing the output equations like so allows that:

$$\boldsymbol{y} = \begin{Bmatrix} z_s \\ z_s - z_u \\ z_u - z_g \end{Bmatrix} \tag{4.117}$$

i.e., the outputs are the sprung mass motion, the suspension travel, and the tire compression. Once the equations are in first order form, the solution can proceed from:

$$\boldsymbol{y}_0 = \mathbf{G}(\mathrm{i}\omega) \boldsymbol{u}_0 \tag{4.118}$$

where:

$$\mathbf{G}(\mathrm{i}\omega) = [\mathbf{C}\ [\mathbf{I}\mathrm{i}\omega - \mathbf{A}]^{-1} \mathbf{B} + \mathbf{D}] \tag{4.119}$$

Figure 4.23 show plots of the resulting equations as functions of frequency, using parameters as given in Table 4.3. Figures 4.24–4.26 repeat the individual

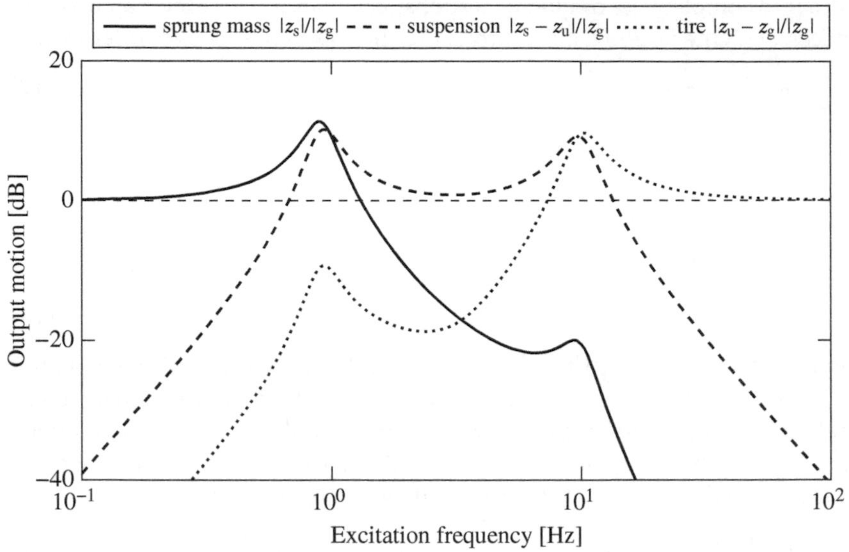

Figure 4.23 The frequency response of the quarter car model shows that at low frequency, disturbances are absorbed by vehicle motion, in the midrange by suspension travel, and at high frequency by tire compression.

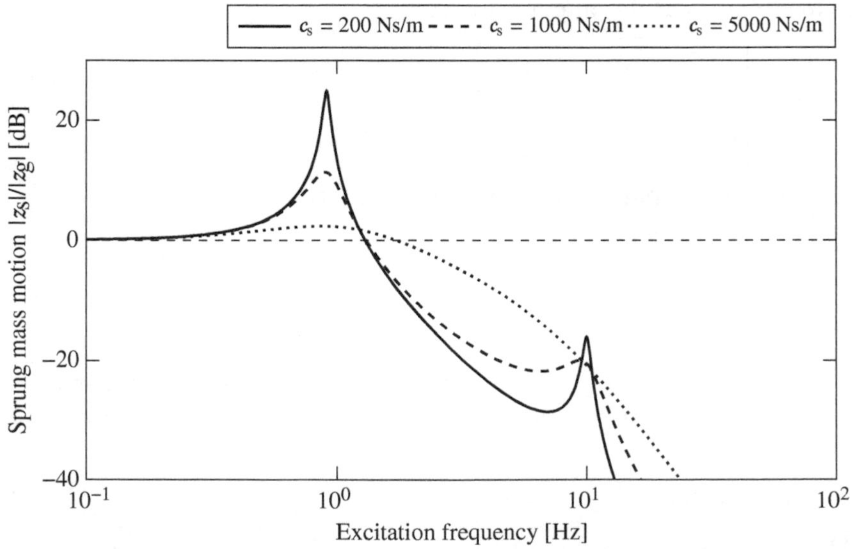

Figure 4.24 The frequency response of the quarter car model shows the fraction of the road disturbance transmitted to the sprung mass. In the low frequency region, the disturbances are amplified. Enough damping to remove the resonance near the natural frequencies improves ride, but too much will decrease midrange performance.

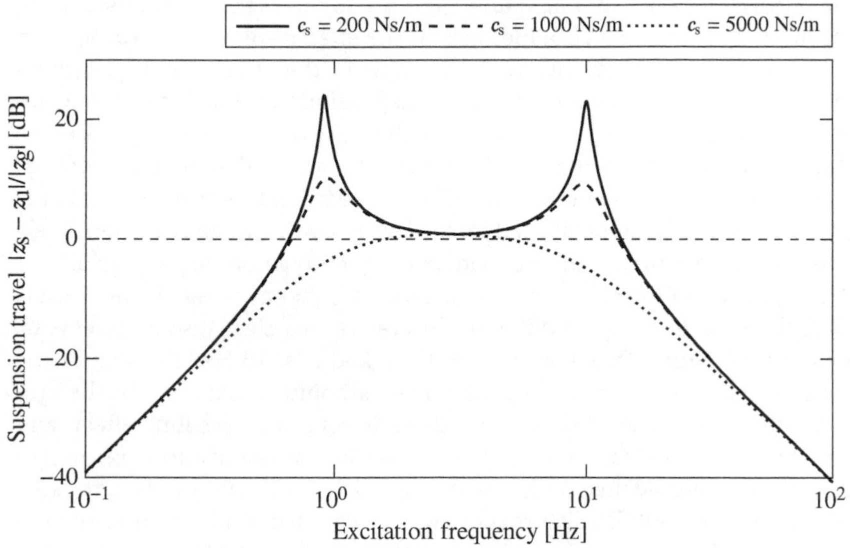

Figure 4.25 The frequency response of the quarter car model shows the fraction of the road disturbance absorbed by suspension travel. The effectiveness is highest in the midrange band of frequencies.

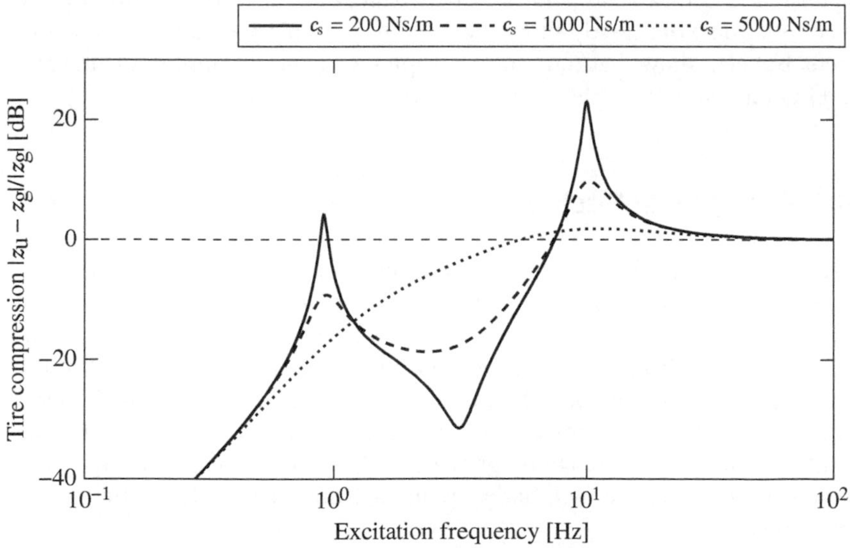

Figure 4.26 The frequency response of the quarter car model shows the fraction of the road disturbance absorbed by tire compression. At high frequencies, the tire absorbs most of the disturbance. As with ride, damping is a compromise. Too much forces the tire to absorb more motion in the midrange, hurting tire grip, while too little results in large resonance effects.

outputs, for a range of damping values. Figure 4.26 shows the ratio of the sprung mass motion to disturbance, a measure of the effectiveness of the suspension in providing ride comfort; the lower the value is, the better it is at absorbing bumps. Figure 4.25 shows the ratio of the suspension travel to disturbance; ideally, this would be 1:1. Figure 4.26 shows the ratio of the tire compression to disturbance, a measure of the effectiveness of the suspension in road handling. This value represents the magnitude of the fluctuation in normal force that is superimposed on the static normal force. The lower this value is, the less likely the tire is to lose contact with the road, consequently improving tire grip.

An inspection of the plots shows that as the frequency of the disturbance is varied, the response is quite different. There are two clear resonance effects, near the two natural frequencies, $\omega \approx 1$ Hz and $\omega \approx 10$ Hz. The size of the resonance effect is highly dependent on the amount of damping in the system. At low frequencies, below 1 Hz, there is very little dynamic effect, and the motion of the sprung mass is almost the same as the disturbance. In this region, the sprung and unsprung masses tend to move in phase. In fact, the suspension actually amplifies disturbances in the region around the lower natural frequency. In the midrange of frequencies, from 1 Hz – 10 Hz, the suspension functions as intended, absorbing the disturbances through relative motion of the sprung and unsprung masses. In this region, the sprung mass will typically move out of phase with the disturbance. At high frequencies, above 10 Hz, the suspension is almost rigid, and the disturbances are absorbed mostly by tire deflection. The plots clearly show the need for damping to control resonance effects, but also show that too much damping will compromise both ride and handling for disturbances between the natural frequencies.

4.4 The Bounce-Pitch Model

As seen in Section 4.3, a study of the quarter car model reveals that the vehicle body motion and the unsprung mass motion tend to occur at very different frequencies. This leads to the idea that the vehicle body motion might be studied independently from the suspension motion. The *bounce-pitch model,* as the name suggests, considers both bounce and pitch motion of the vehicle body, while ignoring any suspension or unsprung mass effects. The assumptions that the pitch angle is small enough to ignore the trigonometry, and that the vehicle can be treated as a single rigid body, are made from the outset. A schematic is shown in Figure 4.27.

The deflection of each of the suspensions can be written in terms of the bounce of the mass centre z and the pitch angle θ. The deflection of the front suspension can be written:

$$z_{\mathrm{f}} = z - a\theta - u_{\mathrm{f}} \tag{4.120}$$

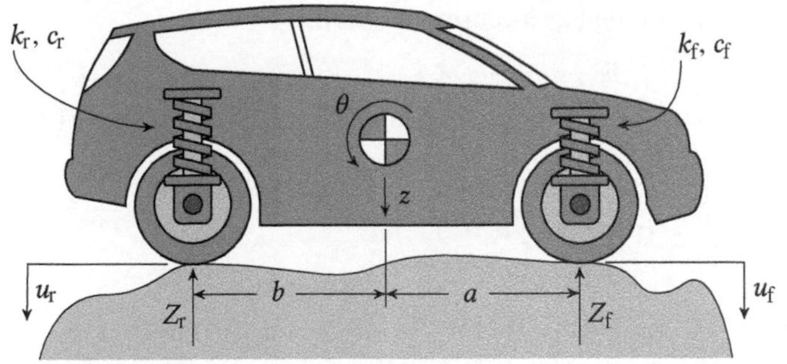

Figure 4.27 The bounce-pitch model can be used to predict ride quality. Note the axis convention and the resulting positive directions.

while at the rear:

$$z_r = z + b\theta - u_r \tag{4.121}$$

Linear expressions including stiffness and damping effects are used to model the suspension forces, which can be summed to give the equation of motion for vehicle bounce.

$$Z_f = k_f z_f + c_f \dot{z}_f \tag{4.122}$$

$$Z_r = k_r z_r + c_r \dot{z}_r \tag{4.123}$$

$$\sum Z = -Z_f - Z_r = m\dot{w} = m\ddot{z} \tag{4.124}$$

$$\begin{aligned} &m\ddot{z} + c_f(\dot{z} - a\dot{\theta}) + c_r(\dot{z} + b\dot{\theta}) + k_f(z - a\theta) + k_r(z + b\theta) \\ &\quad = c_f\dot{u}_f + c_r\dot{u}_r + k_f u_f + k_r u_r \end{aligned} \tag{4.125}$$

Note that when discussing the vertical dynamics the terms c_f and c_r are used to represent the suspension damping and not the tire cornering stiffness. However, the dimensions a and b retain the same meaning: the distance from each axle to the centre of mass. Similarly to the bounce motion, each axle force can be used to find its corresponding pitching moment around the mass centre, which can be summed to give the equation of motion for vehicle pitch:

$$\sum M_G = aZ_f - bZ_r = I_{yy}\ddot{\theta} \tag{4.126}$$

$$\begin{aligned} &I_{yy}\ddot{\theta} - ac_f(\dot{z} - a\dot{\theta}) + bc_r(\dot{z} + b\dot{\theta}) - ak_f(z - a\theta) + bk_r(z + b\theta) \\ &\quad = -ac_f\dot{u}_f + bc_r\dot{u}_r - ak_f u_f + bk_r u_r \end{aligned} \tag{4.127}$$

The two equations are combined as a vector equation:

$$\begin{bmatrix} m_s & 0 \\ 0 & I_{yy} \end{bmatrix} \begin{Bmatrix} \ddot{z} \\ \ddot{\theta} \end{Bmatrix} + \begin{bmatrix} c_f + c_r & bc_r - ac_f \\ bc_r - ac_f & a^2c_f + b^2c_r \end{bmatrix} \begin{Bmatrix} \dot{Z} \\ \dot{\theta} \end{Bmatrix} + \begin{bmatrix} k_f + k_r & bk_r - ak_f \\ bk_r - ak_f & a^2k_f + b^2k_r \end{bmatrix} \begin{Bmatrix} z \\ \theta \end{Bmatrix}$$

$$= \begin{bmatrix} k_f & k_r \\ -ak_f & bk_r \end{bmatrix} \begin{Bmatrix} u_f \\ u_r \end{Bmatrix} + \begin{bmatrix} c_f & c_r \\ -ac_f & bc_r \end{bmatrix} \begin{Bmatrix} \dot{u}_f \\ \dot{u}_r \end{Bmatrix} \tag{4.128}$$

or:

$$\mathbf{M}\ddot{\boldsymbol{z}} + \mathbf{L}\dot{\boldsymbol{z}} + \mathbf{K}\boldsymbol{z} = \mathbf{F}\boldsymbol{u} + \mathbf{G}\dot{\boldsymbol{u}} \tag{4.129}$$

where:

$$\boldsymbol{z} = \begin{Bmatrix} z \\ \theta \end{Bmatrix} \tag{4.130}$$

and the terms in the mass, damping, and stiffness matrices are evident. The form of the equation for the bounce-pitch model is very similar to the quarter car model: a two degree of freedom linear second order ODE. As a result, many of the same types of analysis can be performed.

4.4.1 Transient Analysis

One can compute two natural frequencies for the bounce-pitch model in the same way as for the quarter car model. As before, the damping can be ignored as it will usually be light enough to not have a significant impact on the natural frequencies. The development proceeds identically to give an eigenvector problem.

$$[\mathbf{K} - \omega^2\mathbf{M}]\boldsymbol{z} = \mathbf{0} \tag{4.131}$$

However, unlike the quarter car model, the stiffnesses and mass properties do not result in two widely separate frequencies, nor is there is any equivalent simplification of the characteristic equation.

$$\det \begin{bmatrix} k_f + k_r - m\omega^2 & bk_r - ak_f \\ bk_r - ak_f & a^2k_f + b^2k_r - I_{yy}\omega^2 \end{bmatrix} = 0 \tag{4.132}$$

$$(k_f + k_r - m\omega^2)(a^2k_f + b^2k_r - I_{yy}\omega^2) - (bk_r - ak_f)^2 = 0 \tag{4.133}$$

$$\omega^4 - \left(\frac{k_f + k_r}{m} + \frac{a^2k_f + b^2k_r}{I_{yy}} \right) \omega^2 + \frac{(a+b)^2 k_f k_r}{mI_{yy}} = 0 \tag{4.134}$$

As before, the equation at first appears to be quartic, but is in fact a quadratic in ω^2, and so a solution can be found directly from:

$$\omega^2 = \frac{d_1}{2} \pm \sqrt{\left(\frac{d_1}{2}\right)^2 - d_0} \tag{4.135}$$

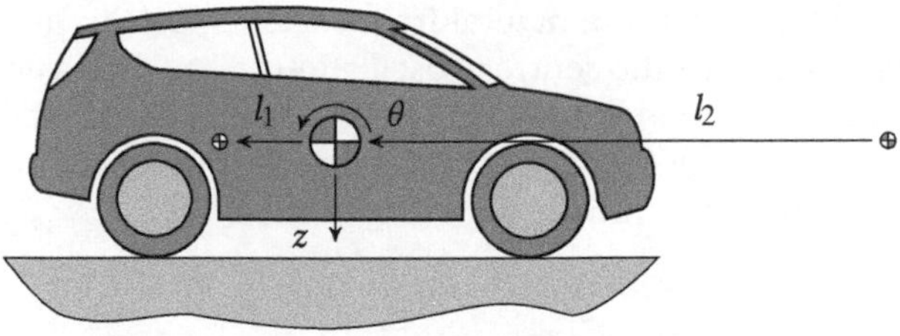

Figure 4.28 The undamped bounce-pitch model has two centres of oscillation that can be found from the eigenvectors. If the model is undamped, the bounce and pitch motions will always be in phase, so a centre of rotation will exist. The total motion will consist of the sum of the two motions around each pitch centre. Ideally the pitch centres lie close to the axles.

where:

$$d_0 = \frac{(a+b)^2 k_f k_r}{m I_{yy}} \tag{4.136}$$

$$d_1 = \frac{k_f + k_r}{m} + \frac{a^2 k_f + b^2 k_r}{I_{yy}} \tag{4.137}$$

In the typical passenger vehicle, the two natural frequencies will both be around 1 Hz. However, there is another difference between the bounce-pitch model and the quarter car. In the quarter car model, each of the natural frequencies are loosely associated with one of the bodies, or with one of the coordinates. In the bounce-pitch model, each of the natural frequencies will contain a certain amount of both bounce and pitch motion. These combined bouncing and pitching motions are described through the definition of the *centre of oscillation*. When the damping is ignored in the model, the bounce and pitch motions must occur exactly in phase, or exactly out of phase. In this case, the combined motion can be considered as a rotation around a fixed centre that is some distance l from the mass centre, as shown in Figure 4.28. This distance can be defined using the relation:

$$z = l\theta \tag{4.138}$$

If the eigenvector solution is enforced, then:

$$(k_f + k_r - m\omega^2)z + (bk_r - ak_f)\theta = 0 \tag{4.139}$$

and:

$$(bk_r - ak_f)z + (a^2 k_f + b^2 k_r - I_{yy}\omega^2)\theta = 0 \tag{4.140}$$

The distance l can then be found as:

$$l = \frac{ak_f - bk_r}{k_f + k_r - m\omega^2} = \frac{a^2 k_f + b^2 k_r - I_{yy}\omega^2}{ak_f - bk_r} \tag{4.141}$$

One value of l can be found for each of the natural frequencies. Typically, one of the values will be small enough that the centre of oscillation will fall between the axles. This indicates that the motion is mostly pitch. The natural frequency associated with this mode is typically the higher of the two. The other centre will land outside the axles, indicating a mostly bouncing motion.

Example

Using the values in Table 4.4, the natural frequencies and oscillation centres can be computed for a typical passenger van using Equations (4.135) – (4.137).

$$\begin{aligned} d_0 &= \frac{(a+b)^2 k_f k_r}{m I_{yy}} \\ &= \frac{(2.885)^2(35)(38) \times 10^6}{(1730)(3267)} \\ &= 1958.6 \end{aligned}$$

$$\begin{aligned} d_1 &= \frac{k_f + k_r}{m} + \frac{a^2 k_f + b^2 k_r}{I_{yy}} \\ &= \frac{(35+38) \times 10^3}{1730} + \frac{(1.189^2(35) + 1.696^2(38)) \times 10^3}{3267} \\ &= 90.799 \end{aligned}$$

$$\omega^2 = \frac{d_1}{2} \pm \sqrt{\left(\frac{d_1}{2}\right)^2 - d_0}$$

Table 4.4 Bounce-pitch model parameter values

Quantity	Notation	Value
front axle dimension	a	1.189 m
rear axle dimension	b	1.696 m
sprung mass	m	1 730 kg
pitch inertia	I_{yy}	3 267 kg m^2
front suspension stiffness	k_f	35 000 N/m
rear suspension stiffness	k_r	38 000 N/m
front suspension damping	c_f	1 000 N/m
rear suspension damping	c_r	1 200 N/m

$$= 45.4 \pm \sqrt{102.56}$$
$$= 35.273, 55.527$$

$$\omega = 5.94 \text{ rad/s}, 7.45 \text{ rad/s}$$
$$= 0.945 \text{ Hz}, 1.19 \text{ Hz}$$

The distance l can then be found as:

$$l = \frac{ak_\mathrm{f} - bk_\mathrm{r}}{k_\mathrm{f} + k_\mathrm{r} - m\omega^2}$$
$$= \frac{(1.189(35) - 1.696(38)) \times 10^3}{(35 + 38) \times 10^3 - (1730)(35.273)}$$
$$= 1.91 \text{ m}$$

and:

$$l = \frac{(1.189(35) - 1.696(38)) \times 10^3}{(35 + 38) \times 10^3 - (1730)(55.527)}$$
$$= -0.990 \text{ m}$$

The two frequencies are around 1 Hz as expected, with the centre of motion for the higher frequency falling inside the wheelbase, indicating more pitch and less bounce motion.

Inspection of this bounce-pitch model shows some interesting special cases, the first of which is when $ak_\mathrm{f} = bk_\mathrm{r}$. Note that this eliminates the off-diagonal terms in the matrix $[\mathbf{K} - \omega^2\mathbf{M}]$ in Equation (4.132), which in turn makes the solution of the eigenvalue problem apparent. In this case, the motions become decoupled. One of the natural frequencies becomes purely pitching motion, when $l = 0$ and:

$$\omega^2 = \frac{a^2k_\mathrm{f} + b^2k_\mathrm{r}}{I_{yy}} \tag{4.142}$$

and one becomes purely bounce, when $l = \infty$ and:

$$\omega^2 = \frac{k_\mathrm{f} + k_\mathrm{r}}{m} \tag{4.143}$$

While mathematically interesting, experience has shown that this configuration generally leads to poor ride quality, and is best avoided if possible.

Another possibility is that $I_{yy} = mab$ (i.e., $k_{yy}^2 = ab$). To explore the effects of this condition, recall the definitions of the axle motions, as in Equations (4.120) and (4.121). Using these definitions, expressions for the bounce and pitch coordinates can be found as:

$$z = \frac{bz_\mathrm{f} + az_\mathrm{r}}{a + b} \tag{4.144}$$

and:

$$\theta = \frac{z_r - z_f}{a+b} \tag{4.145}$$

Substituting these expressions into the equations of motion allows a change of coordinates. The equations of motion become:

$$\frac{1}{a+b}\begin{bmatrix} mb & ma \\ -I_{yy} & I_{yy} \end{bmatrix}\begin{Bmatrix} \ddot{z}_f \\ \ddot{z}_r \end{Bmatrix} + \begin{bmatrix} k_f & k_r \\ -ak_f & bk_r \end{bmatrix}\begin{Bmatrix} z_f \\ z_r \end{Bmatrix} = \begin{Bmatrix} 0 \\ 0 \end{Bmatrix} \tag{4.146}$$

The sprung mass is broken into two fractions, based on the location of the centre of mass; a front mass m_f:

$$m_f = \frac{b}{a+b}m \tag{4.147}$$

and a rear mass m_r:

$$m_r = \frac{a}{a+b}m \tag{4.148}$$

Substituting the inertia condition, and proceeding with the transient analysis gives:

$$\begin{bmatrix} k_f - \omega^2 m_f & k_r - \omega^2 m_r \\ -ak_f + \omega^2 m\frac{ab}{a+b} & bk_r - \omega^2 m\frac{ab}{a+b} \end{bmatrix}\begin{Bmatrix} z_f \\ z_r \end{Bmatrix} = \begin{Bmatrix} 0 \\ 0 \end{Bmatrix} \tag{4.149}$$

or:

$$\begin{bmatrix} k_f - \omega^2 m_f & k_r - \omega^2 m_r \\ -a(k_f - \omega^2 m_f) & b(k_r - \omega^2 m_r) \end{bmatrix}\begin{Bmatrix} z_f \\ z_r \end{Bmatrix} = \begin{Bmatrix} 0 \\ 0 \end{Bmatrix} \tag{4.150}$$

If either the first or the second column can be forced to zeros, then the matrix becomes singular, solving the eigenvalue problem. This occurs when:

$$\omega^2 = \frac{k_f}{m_f} \tag{4.151}$$

or:

$$\omega^2 = \frac{k_r}{m_r} \tag{4.152}$$

The result shows that if the vehicle body's moment of inertia meets the condition $I_{yy} = mab$, then the centres of oscillation land directly on the axles, and the two natural modes correspond to motion of the front suspension only, and motion of the rear suspension only. In the first mode, where $z_r = 0$, or $l = -b$ and the centre of oscillation is on the rear axle, then only the front suspension moves, so its properties alone determine the natural frequency. Conversely, when $z_f = 0$, or $l = a$ and the centre of oscillation is on the front axle, then only the rear suspension properties govern the motion. Unlike the case for bounce and pitch motion, decoupling the front and rear axle motions generally leads to good ride quality. A disturbance at the front axle does not initiate motion at the rear axle, and vice versa.

This result raises the question: what if one were to enforce both of these conditions at once, i.e., $ak_f = bk_r$, and $I_{yy} = mab$? The suspension stiffness and the vehicle body's moment of inertia are physically independent properties, so in theory a vehicle with both conditions could be built. It appears that there is a paradox: how can the centres of oscillation be placed with one at each of the axles, and simultaneously have one at the mass centre and one at infinity? The answer lies in the resulting values of the natural frequency. If $ak_f = bk_r$, then:

$$\omega_1^2 = \frac{k_f + k_r}{m} \tag{4.153}$$

and:

$$\omega_2^2 = \frac{a^2k_f + b^2k_r}{I_{yy}} = \frac{a^2k_f + b^2k_r}{mab} = \frac{\frac{ak_f}{b} + \frac{bk_r}{a}}{m} = \frac{k_r + k_f}{m} \tag{4.154}$$

If both conditions are enforced together, then the two natural frequencies become equal, and the problem is *degenerate*. In the case of repeated eigenvalues, the independence of the eigenvectors is reduced. In the bounce-pitch model under this condition, any combination of bounce and pitch motions will satisfy the eigenvector equations, as both would occur at the same frequency. As might be expected, this generally leads to poor ride quality.

Flat Ride

When selecting the suspension stiffness, even in the case of a front heavy car, where $b > a$, it is common to simultaneously select $k_r > k_f$, i.e., $bk_r > ak_f$. This condition will generally lead to improved ride quality. Experience has shown that most passengers find pitching motions to be more uncomfortable than bounce, so the properties are chosen to discourage pitch. When the vehicle encounters a disturbance, it will generally strike the front axle first, followed by the rear axle a short time later. Setting the rear suspension to have a slightly higher frequency allows its motion to catch up to the front axle; the motions will be more likely to occur in phase, and therefore reduce pitching motion.

Unfortunately, while it is generally quite easy to adjust the stiffness properties of the suspension, the inertia condition is typically much harder to meet. It is a major challenge to significantly change the moment of inertia of the chassis, the location of its centre of mass, or the wheelbase, once the basic vehicle design is complete. As a result, the moment of inertia usually comes close, but does not meet the desired condition exactly.

Maurice Olley was a pioneer in the development of suspensions, particularly in the area of ride quality. His concept of 'flat ride' is well known, and is based on the idea that the suspension should be designed to promote bounce over pitch. Mr Olley summarized the three important design criteria to that effect, often known as the flat ride criteria, as follows:

1. The front suspension must have a greater effective deflection than the rear. Typically front deflection should be at least 30% greater than rear. In other words, the 'spring centre', at which the car, if pushed down vertically, would remain parallel to the ground, should be at least $6^1/_2$% of the wheelbase back of the CG of the sprung mass.
2. The two frequencies of 'pitch' and 'bounce' should be reasonably close together, with a maximum ratio of (pitch frequency/bounce frequency) of 1.20.
3. Ride frequencies should not be greater than 77 cpm, corresponding to an effective ride deflection in a simple system of 6 inches.

Notes:

- The distance from the 'spring centre' to the mass centre of the sprung mass can be found using the expression:

$$c = \frac{bk_r - ak_f}{k_f + k_r} \tag{4.155}$$

  where a positive value of c implies that the spring centre is behind the mass centre.
- The term 'CG' refers to the 'centre of gravity', or equivalently the centre of mass.
- The units cpm are 'cycles per minute'. The value of 77 cpm is equivalent to 1.28 Hz.
- The third criteria utilizes the relationship between static deflection and vibration frequency of a simple sprung mass. Equating the weight to the spring tension caused by deflection Δ gives:

$$k\Delta = mg \rightarrow k = \frac{mg}{\Delta} \tag{4.156}$$

  The natural frequency can then be used to find a relationship to deflection.

$$\omega = \sqrt{\frac{k}{m}} = \sqrt{\frac{mg}{m\Delta}} = \sqrt{\frac{g}{\Delta}} \rightarrow \Delta = \frac{g}{\omega^2} \tag{4.157}$$

Another factor that influences these results is that in practice, the motions will of course be damped, and as a result will not necessarily occur in phase. The natural frequency is found using an eigen analysis, and in the damped case, the eigenvalues and eigenvectors will become complex. The resulting out-of-phase bounce and pitch motions blur the concept of the centre of oscillation, as there is no longer a fixed point of rotation, but rather a travelling wave type of motion that occurs. The calculated distance to the oscillation centre will become a complex number.

If damping is included, it can be shown that if *proportional damping* is used to select the damping values, i.e., $\mathbf{L} = \alpha\mathbf{K} + \beta\mathbf{M}$, for any values of α and β, then the equations of motion can still be decoupled into individual motions. This implies that if the ratio of the damping to the stiffness is the same on both axles, then the bounce and pitch motions will remain in phase, and the centre of oscillation will still exist.

However, yet another complication is that in practice the damping is typically much more nonlinear than the stiffness, usually with less damping when the suspension is in compression than when in rebound, and with the coefficient decreasing with increasing deflection rate.

4.4.2 Frequency Response

The frequency response of the bounce-pitch model describes how the vehicle will respond to disturbances in the road, assuming a sinusoidal input and solution. Unfortunately, the bounce-pitch model is not easily cast into first order form, due to the $\dot{\boldsymbol{u}}$ in the input.

$$\mathbf{M}\ddot{\boldsymbol{z}} + \mathbf{L}\dot{\boldsymbol{z}} + \mathbf{K}\boldsymbol{z} = \mathbf{F}\boldsymbol{u} + \mathbf{G}\dot{\boldsymbol{u}} \tag{4.158}$$

However, with some manipulation, a first order form can be found. First, the motion is broken into two components. The term z_k represents the motion that is the result of the spring force only, and z_c the motion that is the result of the damping force only. Together, they sum to the total motion and solve the equation of motion.

$$\boldsymbol{z} = \boldsymbol{z}_\mathrm{k} + \boldsymbol{z}_\mathrm{c} \tag{4.159}$$

The derivative relation follows directly.

$$\dot{\boldsymbol{z}} = \dot{\boldsymbol{z}}_\mathrm{k} + \dot{\boldsymbol{z}}_\mathrm{c} \tag{4.160}$$

By their definitions, each component solves a modified equation of motion.

$$\mathbf{M}\ddot{\boldsymbol{z}}_\mathrm{k} + \mathbf{K}\boldsymbol{z} = \mathbf{F}\boldsymbol{u} \tag{4.161}$$

$$\mathbf{M}\ddot{\boldsymbol{z}}_\mathrm{c} + \mathbf{L}\dot{\boldsymbol{z}} = \mathbf{G}\dot{\boldsymbol{u}} \tag{4.162}$$

The motion that results from the damping force can be integrated, assuming the effects of the initial conditions will not persist.

$$\mathbf{M}\dot{\boldsymbol{z}}_\mathrm{c} + \mathbf{L}\boldsymbol{z} = \mathbf{G}\boldsymbol{u} \tag{4.163}$$

Finally, the results can be combined to give a standard first order form.

$$\begin{Bmatrix} \dot{z} \\ \ddot{z}_k \end{Bmatrix} = \begin{bmatrix} -\mathbf{M}^{-1}\mathbf{L} & \mathbf{I} \\ -\mathbf{M}^{-1}\mathbf{K} & \mathbf{0} \end{bmatrix} \begin{Bmatrix} z \\ \dot{z}_k \end{Bmatrix} + \begin{bmatrix} \mathbf{M}^{-1}\mathbf{G} \\ \mathbf{M}^{-1}\mathbf{F} \end{bmatrix} u \tag{4.164}$$

$$z = \begin{bmatrix} \mathbf{I} & \mathbf{0} \end{bmatrix} \begin{Bmatrix} z \\ \dot{z}_k \end{Bmatrix} \tag{4.165}$$

Once the equations are in state space form, a standard frequency response to excitation at the front or rear axle can be computed, and is shown in Figure 4.29.

One of the interesting results of the bounce-pitch model is the concept of *wheelbase filtering*. Often the front and rear ground motions are assumed to be the same, only shifted by the time difference required for the vehicle to traverse the bump. The time lag will be:

$$\Delta t = \frac{a+b}{u} \tag{4.166}$$

which, if a sinusoidal road is assumed, corresponds to a phase lag of:

$$\phi = \omega \Delta t = \frac{\omega(a+b)}{u} \tag{4.167}$$

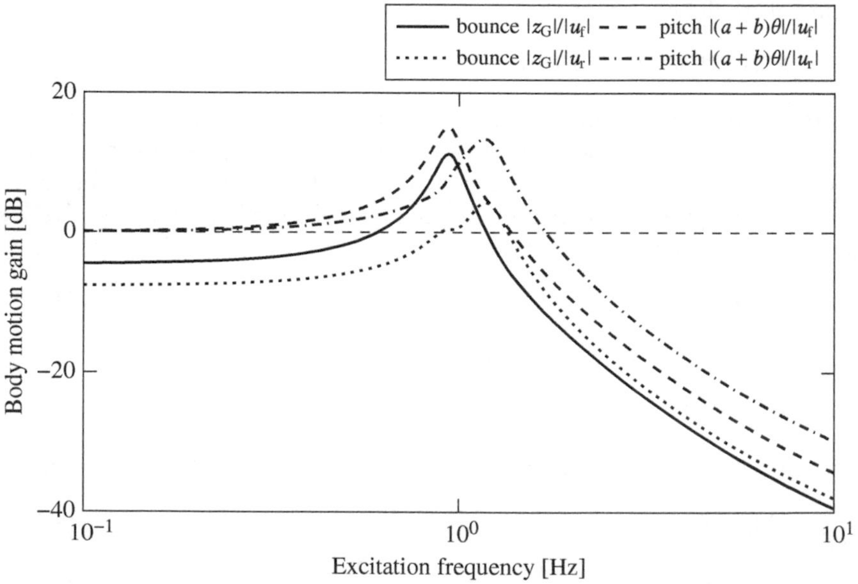

Figure 4.29 The frequency response of the bounce-pitch model can be found using excitation at either the front or rear axle. The responses to front axle excitation show resonances at lower frequencies than those from rear axle excitation, as expected from the different stiffnesses. At very low frequency, the motions tend to those expected from simple geometry, e.g., $\theta = u_f/(a+b)$ and $z_G = u_f b/(a+b)$.

Both the frequency of excitation and the phase lag can be written as functions of the wavelength, the forward speed, and the wheelbase.

$$\omega = \frac{2\pi u}{\lambda} \tag{4.168}$$

$$\phi = \frac{2\pi(a+b)}{\lambda} \tag{4.169}$$

The inputs can be represented mathematically as:

$$\boldsymbol{u} = \begin{Bmatrix} u_{\mathrm{f}} \\ u_{\mathrm{r}} \end{Bmatrix} = \begin{bmatrix} 1 \\ \mathrm{e}^{\mathrm{i}\phi} \end{bmatrix} u_0 \mathrm{e}^{\mathrm{i}\omega t} \tag{4.170}$$

This phase shift causes either the bounce or pitch motions to be alternately excited, based on the wavelength. If the wheelbase happens to be close to the wavelength, or an integer multiple of the wavelength, bounce will be excited. If the wheelbase is close to one half the wavelength, or an odd integer multiple of one half the wavelength, e.g., three halves or five halves, etc., then pitch will be excited. The overall effect is that the wheelbase tends to behave as a filter for certain motions. Figures 4.30–4.32 show the effect of wheelbase filtering over a range of speeds.

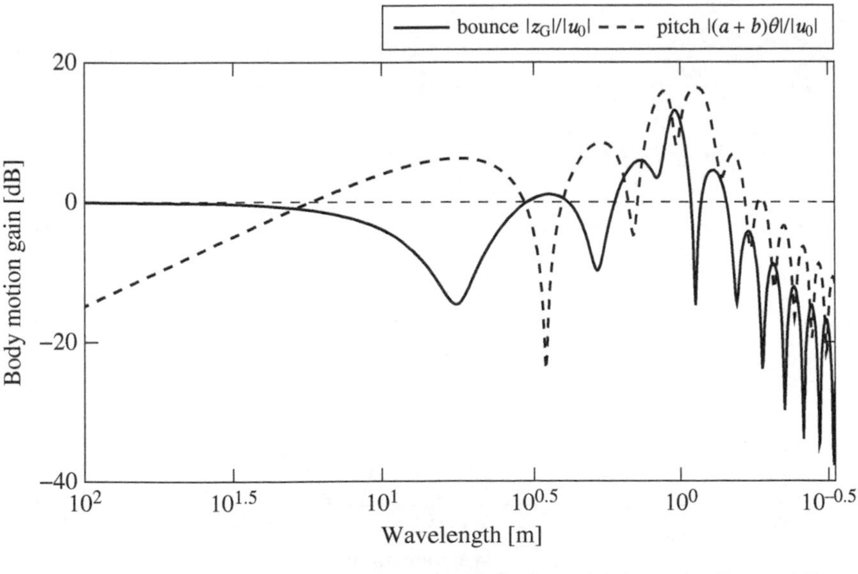

Figure 4.30 The frequency response of the bounce-pitch model shows the filtering effect of the wheelbase. Note that frequency increases with decreasing wavelength, and that at certain wavelengths, only one of the modes is excited, while the other is filtered. The plot show results for a forward speed of $u = 1$ m/s.

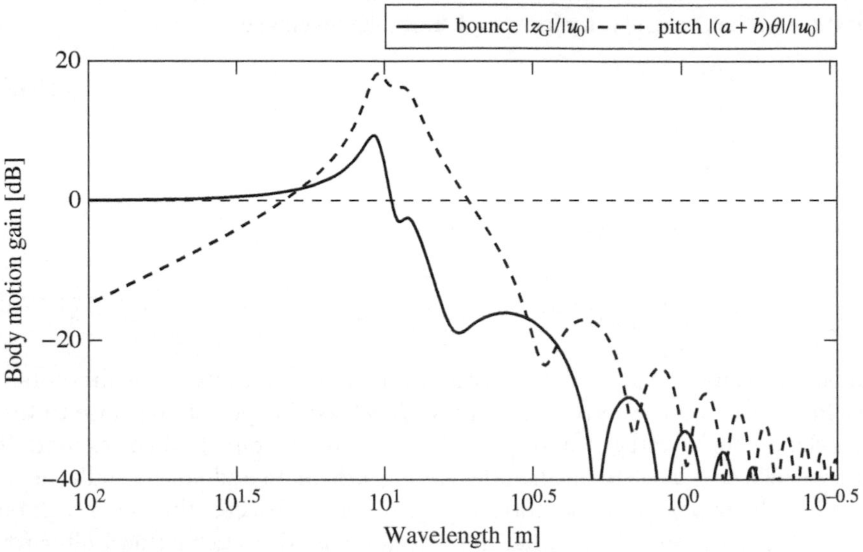

Figure 4.31 The frequency response of the bounce-pitch model shows the filtering effect of the wheelbase at a forward speed of $u = 10$ m/s.

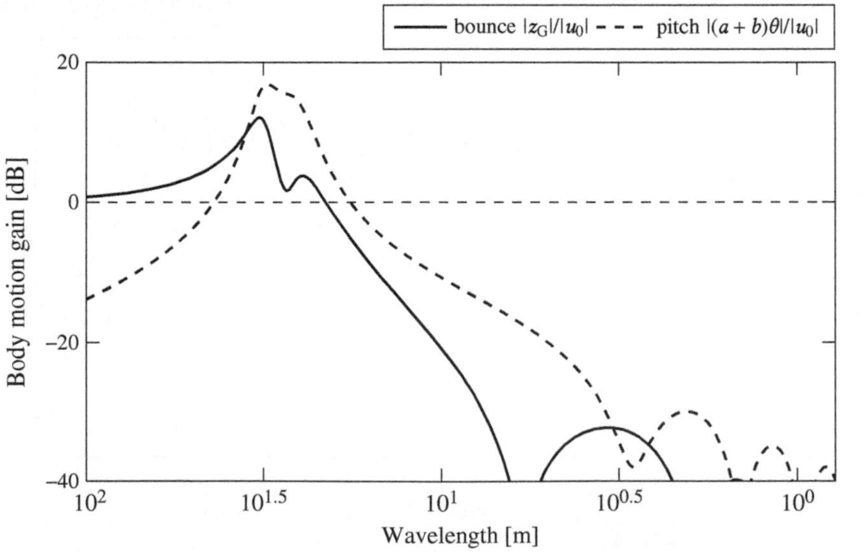

Figure 4.32 The frequency response of the bounce-pitch model shows the filtering effect of the wheelbase at a forward speed of $u = 30$ m/s.

Problems

4.1 Consider a vehicle with the following parameters:
Mass: 2000 kg
Yaw inertia: 3000 kg m^2
Wheelbase ($a + b$): 2.3 m
Weight distribution (%f/r): 55/45
Front tire cornering stiffness c_f: 25 kN/rad (at each front tire)
Rear tire cornering stiffness c_r: 22 kN/rad (at each rear tire)

Assume a linear dynamic model. Suppose that it is lapping on a circular skidpad, at steady state, at 36 km/h, and that it completes a circuit of the skid pad in 20 s.

a) What is the lateral acceleration? Hint: how many radians does the vehicle yaw in one lap?
b) What is the required steer angle?
c) What is the cornering radius?
d) What is the body slip angle?

4.2 Consider the sedan from Problem 4.1.

a) Determine the speed at which the transient handling behaviour will transition to oscillatory, and describe what the eigenvalues tell you about the transient motion at that speed. Hint: the transient solution falls from $\det(\mathbf{M}s + \mathbf{L}) = 0$. Another hint: the roots of a quadratic $as^2 + bs + c = 0$ become complex when $b^2 - 4ac < 0$.
b) Determine the characteristic speed, and describe what the eigenvalues tell you about the transient motion at that speed.
c) Determine the speed at which the rear tires of the vehicle begin to track outside the front. Hint: consider the sign of the steady state transfer function β/δ.

4.3 An episode of a popular UK television show featured a segment on the Citroën 2CV, a small, front engine, front wheel drive, economy car that was mass produced from 1948 until 1990; more than 3.8 million were produced in that span. The 2CV has a four wheel independent suspension that is notably soft, particularly in roll.
The hosts of the show were attempting to test the rumor that the 2CV was impossible to roll over. Despite their best attempts and numerous types of cornering manœuvres, at both low and high speed, the correspondents were unable to cause the 2CV to roll. It demonstrated a distinct understeer behaviour. Finally, one of the correspondents put the 2CV in reverse, backed up moderately quickly, steered hard, and promptly rolled the car onto its roof.

a) Why do you think this would happen, and can you predict this result with the bicycle model equations? How?
b) The bicycle model equations for a car with rear wheel steering are given below. How do you think the steady state transfer functions might compare for a car with front steering?

$$\begin{bmatrix} m & 0 \\ 0 & I_{zz} \end{bmatrix} \begin{Bmatrix} \dot{v} \\ \dot{r} \end{Bmatrix} + \frac{1}{u} \begin{bmatrix} c_f + c_r & ac_f - bc_r + mu^2 \\ ac_f - bc_r & a^2 c_f + b^2 c_r \end{bmatrix} \begin{Bmatrix} v \\ r \end{Bmatrix} = \begin{bmatrix} c_f & c_r \\ ac_f & -bc_r \end{bmatrix} \begin{Bmatrix} \delta_f \\ \delta_r \end{Bmatrix} \tag{4.171}$$

4.4 Consider the system shown in Problem 4.4: a truck towing a trailer, which in turn is towing another trailer. The system can be modelled with four degrees of freedom, namely lateral motion of the truck, and yaw of the truck and each trailer. The system can be treated as having a constant forward speed.

a) How many eigenvalues would such a system have, and how many potential natural frequencies?
b) Sketch and describe the mode shapes that you would expect the eigenvectors of the system to indicate at medium to highway speed.

4.5 Consider a vehicle with the following specifications:
Sprung mass: 1250 kg
Pitch moment of inertia: 1500 kg m^2
Wheelbase: 2.3 m
Distance from the centre of mass to the front axle: 0.8 m
Front suspension stiffness k_f: 15 kN/m
Rear suspension stiffness k_r: 18 kN/m
Front unsprung mass m_{uf}: 20 kg

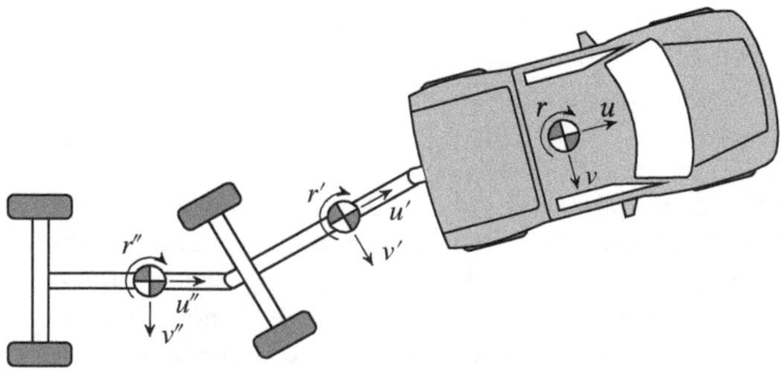

Problem 4.4 Double trailer system

Rear unsprung mass m_{ur}: 16 kg
Tire vertical stiffness k_t: 160 kN/m

Note that the figures quoted for the suspension stiffness, unsprung mass, and tire stiffness are for a single wheel and tire only.

a) Construct a quarter car model for each of the front and rear suspensions, and find the resulting frequencies. Comment on the results. Are they close to what you expect?
b) Compare the results to the frequencies predicted by the bounce-pitch model, but first, modify the suspension stiffnesses to account for the tire by using the effective stiffness, where $\frac{1}{k_{eff}} = \frac{1}{k_t} + \frac{1}{k_s}$. (Note that the effective stiffness of two springs in series are less than the stiffness of the least stiff spring individually.) What do you see from the result?

4.6 During the 2006 Formula 1 season, there was considerable controversy over the use of *mass damper* systems, also known as *vibration absorbers*, and the subsequent banning of these devices by the FIA. Vibration absorbers are spring-mass-damper systems that work by setting the natural frequency of the absorber to the frequency of the motion that one wishes to cancel, in this case, the chassis motion. The cancelled motion is instead transferred to the absorber. A schematic diagram of the system is shown in Problem 4.6.

a) Suppose you were hoping to tune the natural frequency of the damper to the bounce frequency of the chassis. Recognizing that the wheel rate on a modern Formula 1 car would be much higher than a typical passenger car, on the order of 100000 N/m, explain why this would affect your solution.
b) Derive the equations of motion of the quarter car model of the vehicle, including the mass damper system. What is the most significant effect of adding the mass damper on the equations of motion?
c) Describe how you might use the equations above to help tune the vibration absorber.

4.7 An *inerter* is a device that is sometime used in vehicle suspensions in special applications. The device is similar to a spring or damper, in that it connects the sprung and unsprung mass in the suspension. However, unlike a spring, which depends on relative displacement, or a damper, which depends on relative velocity, the force in an inerter depends on the *relative acceleration* of the ends of the device. Physically, in simple terms, it includes internal rotating components. Compressing or extending the inerter requires these components to rotate quickly, such that their rotational inertia is the dominant factor in the motion. As a result, the effective mass is much larger than the actual mass. A schematic diagram of the system is shown in Problem 4.6. The force in the inerter can be modelled with the expression:

$$f_i = n(\ddot{z}_u - \ddot{z}_s)$$

where z_s is the sprung mass displacement, z_u is the unsprung mass displacement, and n is a constant coefficient. A positive force indicates the device is in tension.

a) Draw free body diagrams of the sprung and unsprung mass of a quarter car model, with a spring, a damper, and an inerter included in the suspension. Write the equations of motion, using z_s and z_u as the coordinates. Do not attempt to reduce the equations to first order form.
b) Suppose that the unsprung mass $m_u = 30$ kg, the sprung mass $m_s = 400$ kg, the suspension stiffness $k_s = 20000$ N/m, the tire stiffness $k_t = 120000$ N/m, and $n = 20$ N s^2/m. Solve for the undamped natural frequencies of the model. Hint: the approximate quarter car formulae no longer apply.

4.8 Consider a vehicle with the following specifications:
Mass: 3000 kg
Wheelbase: 3.55 m
Weight distribution (%f/r): 58/42

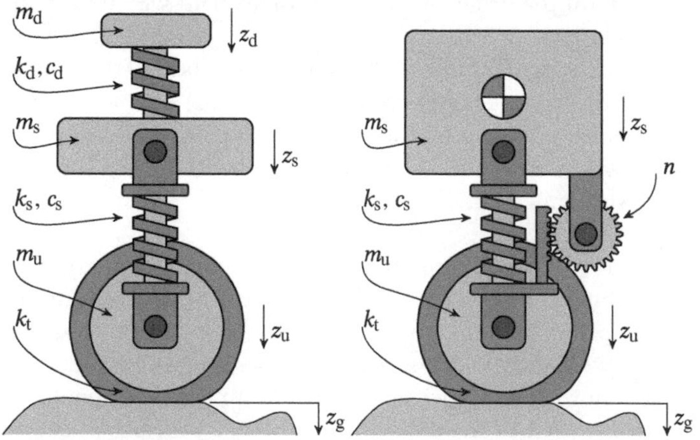

(a) quarter car with mass damper (b) quarter car with inerter

Problem 4.6 Variations on the quarter car model

A simple undamped bounce-pitch model of this vehicle has been constructed, and an eigenvector analysis of the equations of motion yields the following result:

$$\begin{Bmatrix} z_0 \\ \theta_0 \end{Bmatrix}_1 = \begin{Bmatrix} 1.0 \\ -0.486 \end{Bmatrix}, \begin{Bmatrix} z_0 \\ \theta_0 \end{Bmatrix}_2 = \begin{Bmatrix} 1.0 \\ 0.671 \end{Bmatrix}$$

a) Describe the physical significance of the results. Are they reasonable?
b) Can you predict anything about the ride quality from this information alone? If so, what does it indicate?
c) Suppose that you are also given the information that $\omega_1 = 7.54$ rad/s and $\omega_2 = 8.16$ rad/s. Does this give any other indicators of ride quality? Can you determine the suspension stiffnesses from this information?

4.9 Consider a vehicle with the following specifications:
Sprung mass: 1550 kg
Pitch moment of inertia: 1850 kg m^2
Wheelbase: 2.3 m
Distance from the centre of mass to the front axle: 1.0 m
Front suspension stiffness k_f: 18 kN/m (at each front wheel)
Rear suspension stiffness k_r: 22 kN/m (at each rear wheel)

Assume a linear dynamic model.

a) Construct a bounce-pitch model, and find the resulting frequencies.
b) Find the location of the instant centres of oscillation.
c) Using the criteria of Maurice Olley, comment on the predicted ride quality of the vehicle. Do you have any suggestions to improve it?

4.10 Suppose you are requested to conduct a ride quality analysis on a vehicle with the following properties:
Sprung mass: 3000 kg
Pitch moment of inertia: 4500 kg m^2
Wheelbase: 3.55 m
Weight distribution (%f/r): 58/42

A simple undamped bounce-pitch model using a typical coordinate system of vertical deflection of the centre of mass, and the pitch angle has been developed. An eigenvector analysis of the equations of motion yields the following result:

$$\begin{Bmatrix} z_0 \\ \theta_0 \end{Bmatrix}_1 = \begin{Bmatrix} 1.0 \\ 0 \end{Bmatrix}, \begin{Bmatrix} z_0 \\ \theta_0 \end{Bmatrix}_2 = \begin{Bmatrix} 0 \\ 1.0 \end{Bmatrix}$$

a) Describe the physical significance of the results. Can you predict anything about the ride quality from this information alone? If so, what does it indicate?
b) Suppose that you are also given the information that $k_f = 35000$ N/m. Can you determine any other relevant information about the ride behaviour?

4.11 Given a passenger car with the following parameters:
Mass: 2000 kg
Pitch moment of inertia: 2600 kg m^2
Roll moment of inertia: 1200 kg m^2
Front suspension stiffness: 22 kN/m (*at each wheel*)
Rear suspension stiffness: 18 kN/m (*at each wheel*)
Anti-roll bar stiffness: front only, 5 kN/m (measured by deflecting one suspension while holding the other side fixed, this is in addition to the normal suspension stiffness)
Wheelbase: 2.3 m
Distance from the centre of mass to the front axle: 1.1 m
Height of the centre of mass: 0.6 m (above the ground)
Track width: 1.4 m (left to right, same for both front and rear axles)

a) Develop a bounce-pitch model of the vehicle. Determine the natural frequencies, and locate the oscillation centres. Describe the motion in each of the two modes.
b) Consider the analogous model for bounce and roll, and find the roll natural frequency. Hint: there is usually a coupling between bounce and pitch, but roll will be decoupled. (The off-diagonal terms of the stiffness matrix are eliminated by symmetry.) Start by finding the roll stiffness, by writing expressions for the tire normal forces as a function of roll. Think carefully about the effect of the anti-roll bar (use a simple linear relationship).
c) Discuss the results in light of the ride quality criteria of Maurice Olley.

References

1 Rill, G., 2011. *Road Vehicle Dynamics: Fundamentals and Modeling*. CRC Press.

5

Full Car Model

Another useful model in the study of vehicle dynamics is the full car model. A number of variations on the model are possible. Generally, the bounce, pitch, and roll motions of the vehicle body are considered. The response to steady cornering on a smooth road is of interest, as is the response to a rough surface.

5.1 Steady State Analysis

The first variation presented is a *quasi-static* model, i.e., the accelerations are assumed to be constant, so that loads and suspension deflections will also be constant, and the vehicle is treated as a single rigid body. The development starts with the general equations of motion, but will initially focus on three of these equations, namely, bounce, roll, and pitch.

$$\sum \boldsymbol{f} = m\boldsymbol{a}_{\mathrm{G}} = m(\dot{\boldsymbol{v}} + \boldsymbol{\omega} \times \boldsymbol{v}) \tag{5.1}$$

or:

$$\sum X\hat{\mathbf{i}} + Y\hat{\mathbf{j}} + Z\hat{\mathbf{k}} = m((\dot{u}\hat{\mathbf{i}} + \dot{v}\hat{\mathbf{j}} + \dot{w}\hat{\mathbf{k}}) + (p\hat{\mathbf{i}} + q\hat{\mathbf{j}} + r\hat{\mathbf{k}}) \times (u\hat{\mathbf{i}} + v\hat{\mathbf{j}} + w\hat{\mathbf{k}})) \tag{5.2}$$

5.1.1 The Bounce-Pitch-Roll Model

The longitudinal equation is not relevant, and ignored. The lateral equation is skipped for the time being, and will be considered after the expressions for bounce, roll, and pitch have been prepared. When considering the vertical forces, one should notice that the problem is *statically indeterminate* (or perhaps *quasi-statically indeterminate*), i.e., there are four unknown forces, but only three useful equations. These must be supplemented with equations describing the stiffness of the system. When writing the vertical equation, the

Fundamentals of Vehicle Dynamics and Modelling: A Textbook for Engineers with Illustrations and Examples, First Edition. Bruce P. Minaker.

Companion website: www.wiley.com/go/minaker/vehicle-dynamics

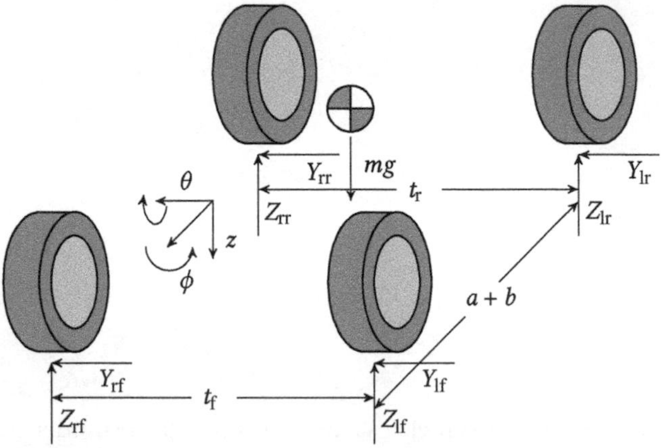

Figure 5.1 The free body diagram of the full car model has vertical and lateral forces acting at each of the tire contact patches.

roll rate and pitch rate are assumed to be zero, and the vehicle is assumed to be in vertical equilibrium. A free body diagram is shown in Figure 5.1.

$$\sum Z = -Z_{\mathrm{rf}} - Z_{\mathrm{lf}} - Z_{\mathrm{rr}} - Z_{\mathrm{lr}} + mg = m\dot{w} + pv - uq = 0 \tag{5.3}$$

The general equation for rotation is considered next.

$$\sum \boldsymbol{m}_{\mathrm{G}} = \mathbf{I}_{\mathrm{G}}\boldsymbol{\alpha} + \boldsymbol{\omega} \times \mathbf{I}_{\mathrm{G}}\boldsymbol{\omega} \tag{5.4}$$

The left-right symmetry of the vehicle is used to eliminate the cross products of inertia I_{xy} and I_{yz}. In the convention used here, a positive I_{xz} implies that the rear of the vehicle carries its mass further above the ground than the front.[1]

$$\begin{aligned}
\begin{Bmatrix} L \\ M \\ N \end{Bmatrix} &= \begin{bmatrix} I_{\mathrm{xx}} & -I_{\mathrm{xy}} & -I_{\mathrm{xz}} \\ -I_{\mathrm{xy}} & I_{\mathrm{yy}} & -I_{\mathrm{yz}} \\ -I_{\mathrm{xz}} & -I_{\mathrm{yz}} & I_{\mathrm{zz}} \end{bmatrix} \begin{Bmatrix} \dot{p} \\ \dot{q} \\ \dot{r} \end{Bmatrix} + (p\hat{\mathbf{i}} + q\hat{\mathbf{j}} + r\hat{\mathbf{k}}) \times \begin{bmatrix} I_{\mathrm{xx}} & -I_{\mathrm{xy}} & -I_{\mathrm{xz}} \\ -I_{\mathrm{xy}} & I_{\mathrm{yy}} & -I_{\mathrm{yz}} \\ -I_{\mathrm{xz}} & -I_{\mathrm{yz}} & I_{\mathrm{zz}} \end{bmatrix} \begin{Bmatrix} p \\ q \\ r \end{Bmatrix} \\
&= \begin{bmatrix} I_{\mathrm{xx}} & 0 & -I_{\mathrm{xz}} \\ 0 & I_{\mathrm{yy}} & 0 \\ -I_{\mathrm{xz}} & 0 & I_{\mathrm{zz}} \end{bmatrix} \begin{Bmatrix} \dot{p} \\ \dot{q} \\ \dot{r} \end{Bmatrix} + (p\hat{\mathbf{i}} + q\hat{\mathbf{j}} + r\hat{\mathbf{k}}) \times \begin{bmatrix} I_{\mathrm{xx}} & 0 & -I_{\mathrm{xz}} \\ 0 & I_{\mathrm{yy}} & 0 \\ -I_{\mathrm{xz}} & 0 & I_{\mathrm{zz}} \end{bmatrix} \begin{Bmatrix} p \\ q \\ r \end{Bmatrix}
\end{aligned} \tag{5.5}$$

---

1 In the SAE sign convention, positive x and z are forward and down, respectively. As a result, mass that is located forward and low, or rearward and high contributes to a positive cross product under this definition. The off-diagonal entries in the inertia matrix are taken as $-I_{\mathrm{xz}}$, etc., and the definition of I_{xz} is the positive value of the integral, i.e., $I_{\mathrm{xz}} = \int_m xz\, dm$. Some authors choose to include the negative sign in the definition of the I_{xz} term itself rather than in the inertia matrix, often leading to confusion about the signs of the cross products.

These collapse to:

$$\sum L = I_{xx}\dot{p} - (I_{yy} - I_{zz})qr - I_{xz}(\dot{r} + pq) \tag{5.6}$$

$$\sum M = I_{yy}\dot{q} - (I_{zz} - I_{xx})pr - I_{xz}(r^2 - p^2) \tag{5.7}$$

$$\sum N = I_{zz}\dot{r} - (I_{xx} - I_{yy})pq - I_{xz}(\dot{p} - qr) \tag{5.8}$$

It must be noted that the equations as written assume that the moments are summed around the centre of mass G. If another reference point A is chosen, then the equations must be modified to include this effect.

$$\sum m_A = \sum m_G + \mathbf{r}_{G/A} \times m\mathbf{a}_G \tag{5.9}$$

The roll moment equation is written using an axis in the ground plane as the reference. This eliminates moments due to lateral tire forces, but does require the addition of the inertial moment due to the nonzero distance between the reference point and the centre of mass. The roll and yaw angular accelerations are assumed to be zero in the steady state, as is the pitch rate. As a result, the effect of the cross product I_{xz} is eliminated. Of course the centre of mass is assumed to be above the ground, but one should note that due to the SAE sign convention, this is in the negative z direction relative to the reference point in the ground plane.

$$\sum L\hat{\mathbf{i}} = 0 - 0 - 0 + (-h_G\hat{\mathbf{k}} \times mur\hat{\mathbf{j}}) = murh_G\hat{\mathbf{i}} \tag{5.10}$$

or:

$$\sum L = \frac{t_f}{2}(Z_{lf} - Z_{rf}) + \frac{t_r}{2}(Z_{lr} - Z_{rr}) = murh_G \tag{5.11}$$

In the pitch direction, the pitch angular acceleration is assumed to be zero, and the forward velocity is assumed to be constant, so the vertical offset of the reference from the centre of mass does not add any terms. However, an inertial effect due to the cross product I_{xz} is present. This term is nonzero if the weight at either the front or rear of the vehicle is significantly higher than at the other end. This configuration when combined with a yaw rotational velocity will produce an inertial pitching moment that tends to lift the front of the vehicle. The effect is usually small.

$$\sum M = a(Z_{rf} + Z_{lf}) - b(Z_{rr} + Z_{lr}) = I_{yy}\dot{q} - I_{xz}r^2 = -I_{xz}r^2 \tag{5.12}$$

To form a solvable set of equations, expressions for the normal forces are required. The assumption is made that the normal forces are linear in terms of the suspension compression, with coupling across each axle due to the anti-roll bar. The effect of the tire stiffness is neglected, as it is much higher than the suspension stiffness, and has a relatively small effect on the overall force. (Recall that the stiffness of two springs in series is less than the stiffness of either of the individual springs.) Note that the anti-roll bar stiffness used here is given as a linear stiffness, measured as the force applied at the wheel sufficient to cause a unit of relative deflection between the left and right sides.[2]

$$Z_{\mathrm{rf}} = k_{\mathrm{f}} z_{\mathrm{rf}} + k_{\mathrm{fb}}(z_{\mathrm{rf}} - z_{\mathrm{lf}}) \tag{5.13}$$

$$Z_{\mathrm{lf}} = k_{\mathrm{f}} z_{\mathrm{lf}} + k_{\mathrm{fb}}(z_{\mathrm{lf}} - z_{\mathrm{rf}}) \tag{5.14}$$

$$Z_{\mathrm{rr}} = k_{\mathrm{r}} z_{\mathrm{rr}} + k_{\mathrm{rb}}(z_{\mathrm{rr}} - z_{\mathrm{lr}}) \tag{5.15}$$

$$Z_{\mathrm{lr}} = k_{\mathrm{r}} z_{\mathrm{lr}} + k_{\mathrm{rb}}(z_{\mathrm{lr}} - z_{\mathrm{rr}}) \tag{5.16}$$

Expressions for the suspension deflection can be written in terms of the bounce, pitch, and roll motions.

$$\begin{Bmatrix} z_{\mathrm{rf}} \\ z_{\mathrm{lf}} \\ z_{\mathrm{rr}} \\ z_{\mathrm{lr}} \end{Bmatrix} = \begin{bmatrix} 1 & \frac{t_{\mathrm{f}}}{2} & -a \\ 1 & -\frac{t_{\mathrm{f}}}{2} & -a \\ 1 & \frac{t_{\mathrm{r}}}{2} & b \\ 1 & -\frac{t_{\mathrm{r}}}{2} & b \end{bmatrix} \begin{Bmatrix} z \\ \phi \\ \theta \end{Bmatrix} \tag{5.17}$$

Assembling the vertical, roll, and pitch equations, along with the normal force expressions, into one set gives:

$$\begin{bmatrix} 1 & 1 & 1 & 1 & 0 & 0 & 0 \\ -t_{\mathrm{f}} & t_{\mathrm{f}} & -t_{\mathrm{r}} & t_{\mathrm{r}} & 0 & 0 & 0 \\ a & a & -b & -b & 0 & 0 & 0 \\ -1 & 0 & 0 & 0 & k_{\mathrm{f}} & \frac{1}{2}t_{\mathrm{f}}k_{\mathrm{fe}} & -ak_{\mathrm{f}} \\ 0 & -1 & 0 & 0 & k_{\mathrm{f}} & -\frac{1}{2}t_{\mathrm{f}}k_{\mathrm{fe}} & -ak_{\mathrm{f}} \\ 0 & 0 & -1 & 0 & k_{\mathrm{r}} & \frac{1}{2}t_{\mathrm{r}}k_{\mathrm{re}} & bk_{\mathrm{r}} \\ 0 & 0 & 0 & -1 & k_{\mathrm{r}} & -\frac{1}{2}t_{\mathrm{r}}k_{\mathrm{re}} & bk_{\mathrm{r}} \end{bmatrix} \begin{Bmatrix} Z_{\mathrm{rf}} \\ Z_{\mathrm{lf}} \\ Z_{\mathrm{rr}} \\ Z_{\mathrm{lr}} \\ z \\ \phi \\ \theta \end{Bmatrix} = \begin{Bmatrix} mg \\ 2murh_{\mathrm{G}} \\ -I_{xz}r^2 \\ 0 \\ 0 \\ 0 \\ 0 \end{Bmatrix} \tag{5.18}$$

---

2 The roll stiffness is often quoted as the moment per unit rotation of the sprung mass; the conversion factor is the square of the track width (e.g., in metric units, k_{fb} [Nm/rad] $= t^2 k_{\mathrm{fb}}$ [N/m]).

where:

$$k_{fe} = k_f + 2k_{fb} \tag{5.19}$$

$$k_{re} = k_r + 2k_{rb} \tag{5.20}$$

With some considerable algebra, the set of equations can be solved. It is noteworthy that all four normal forces can be found independently of the lateral tire forces. The result shows two components, one for static load, and one for weight transfer.

$$Z_{rf} = \underbrace{\frac{mg}{2}\frac{b}{a+b}}_{\text{static load}} \underbrace{- \frac{I_{xz}r^2}{2(a+b)} - murh_G\left(\frac{t_f k_{fe}}{t_f^2 k_{fe} + t_r^2 k_{re}}\right)}_{\text{weight transfer}} \tag{5.21}$$

The static load is simply the appropriate fraction of the weight carried by each corner of the vehicle. The weight transfer has two components; the first is the result of the inertial pitching moment lifting the front of the vehicle, and the second is the rolling moment times the roll stiffness fraction of the individual axle. Considering the case where the front and rear track width are the same (i.e., $t_f = t_r = t$) shows the familiar lateral weight transfer term.

$$Z_{rf} = \frac{mg}{2}\frac{b}{a+b} - \frac{I_{xz}r^2}{2(a+b)} - \frac{murh_G}{t}\left(\frac{k_{fe}}{k_{fe} + k_{re}}\right) \tag{5.22}$$

The expressions for the remaining forces follow from symmetry.

$$Z_{lf} = \frac{mgb - I_{xz}r^2}{2(a+b)} + murh_G\left(\frac{t_f k_{fe}}{t_f^2 k_{fe} + t_r^2 k_{re}}\right) \tag{5.23}$$

$$Z_{rr} = \frac{mga + I_{xz}r^2}{2(a+b)} - murh_G\left(\frac{t_f k_{fe}}{t_f^2 k_{fe} + t_r^2 k_{re}}\right) \tag{5.24}$$

$$Z_{lr} = \frac{mga + I_{xz}r^2}{2(a+b)} + murh_G\left(\frac{t_f k_{fe}}{t_f^2 k_{fe} + t_r^2 k_{re}}\right) \tag{5.25}$$

A simple expression for the roll angle can also be found. The roll is a function of the lateral acceleration, and not the static weight or yaw rate.

$$\phi = \frac{2murh_G}{t_f^2 k_{fe} + t_r^2 k_{re}} \tag{5.26}$$

As one might expect, the expression is equal to the roll moment over the effective roll stiffness. Again, considering the case where the front and rear track width are the same gives:

$$\phi = \frac{2murh_G}{t^2(k_{fe} + k_{re})} = \frac{murh_G}{\frac{t^2}{2}(k_f + k_r) + t^2(k_{fb} + k_{rb})} \tag{5.27}$$

Finally, expressions for the bounce and pitch motion can also be found.

$$z = \frac{I_{xz} r^2 (a k_f - b k_r) + mg(a^2 k_f + b^2 k_r)}{2 k_f k_r (a + b)^2} \tag{5.28}$$

$$\theta = \frac{I_{xz} r^2 (k_f + k_r) + mg(a k_f - b k_r)}{2 k_f k_r (a + b)^2} \tag{5.29}$$

Note that neither bounce nor pitch depend on lateral acceleration. Both expressions contain only the static deflection term due to weight, and a term in response to the front to rear weight transfer caused by inertial pitching moment. In most cases, the deflection due to the inertial pitching moment is small and can usually be safely ignored. Additionally, the bounce and pitch motions are typically measured from the equilibrium rest position rather than the zero suspension deflection position. As a result, the remaining portions of the expressions can best be thought of as the amount of lift and pitch necessary to lift all four tires just clear of the ground. However, in practice, most road vehicles have stops built into the suspensions to limit the amount of travel to prevent the springs from ever coming slack in their mounts. The actual amount of lift and pitch required would be less than the results of the calculation.

5.1.2 The Lateral-Yaw Model

The lateral force equation is found by considering the y component of the equation of motion. If one assumes that the vehicle is cornering in steady state, i.e. $\dot{v} = p = 0$, then the lateral acceleration simplifies to just ur.

$$\sum Y = Y_{rf} + Y_{lf} + Y_{rr} + Y_{lr} = m(\dot{v} + ru - wp) = mur \tag{5.30}$$

The yaw equation, assuming steady state with zero roll rate and pitch rate, is:

$$\sum N = a(Y_{rf} + Y_{lf}) - b(Y_{rr} + Y_{lr}) = I_{zz}\dot{r} - (I_{xx} - I_{yy})pq - I_{xz}(\dot{p} - qr) = 0 \tag{5.31}$$

Using these two equations, the total (left plus right) lateral force can be found at each axle. Let $Y_{rf} + Y_{lf} = Y_f$, and $Y_{rr} + Y_{lr} = Y_r$. This implies:

$$Y_f = \frac{b}{a + b} mur \tag{5.32}$$

and:

$$Y_r = \frac{a}{a + b} mur \tag{5.33}$$

Now, if an appropriate tire model is used, along with some assumptions on slip angle, the individual tire lateral forces can be found. For example, suppose the lateral tire force could be expressed as some function of slip, times a parabolic function of normal load, such as:

$$Y_{rf} = f(\alpha_{rf})(\mu Z_{rf} - \kappa Z_{rf}^2) \tag{5.34}$$

where μ represents the coefficient of friction, and κ is the rate at which μ decreases with increasing normal load. In this case, if the tire slip angle is assumed to be the same on both front tires, i.e., $\alpha_{rf} = \alpha_{lf} = \alpha_f$, and $Z_{rf} + Z_{lf} = Z_f$ and $Z_{rr} + Z_{lr} = Z_r$, then:

$$\begin{aligned} Y_f &= f(\alpha_{rf})(\mu Z_{rf} - \kappa Z_{rf}^2) + f(\alpha_{lf})(\mu Z_{lf} - \kappa Z_{lf}^2) \\ &= f(\alpha_f)(\mu Z_{rf} - \kappa Z_{rf}^2 + \mu Z_{lf} - \kappa Z_{lf}^2) \\ &= f(\alpha_f)(\mu Z_f - \kappa(Z_{rf}^2 + Z_{lf}^2) \end{aligned} \tag{5.35}$$

By substitution:

$$f(\alpha_f) = \frac{b}{a+b}\frac{mur}{\mu Z_f - \kappa(Z_{rf}^2 + Z_{lf}^2)} \tag{5.36}$$

or:

$$Y_{rf} = \frac{b}{a+b}\frac{\mu Z_{rf} - \kappa Z_{rf}^2}{\mu Z_f - \kappa(Z_{rf}^2 + Z_{lf}^2)} mur \tag{5.37}$$

and by symmetry:

$$Y_{lf} = \frac{b}{a+b}\frac{\mu Z_{lf} - \kappa Z_{lf}^2}{\mu Z_f - \kappa(Z_{rf}^2 + Z_{lf}^2)} mur \tag{5.38}$$

$$Y_{rr} = \frac{a}{a+b}\frac{\mu Z_{rr} - \kappa Z_{rr}^2}{\mu Z_r - \kappa(Z_{rr}^2 + Z_{lr}^2)} mur \tag{5.39}$$

$$Y_{lr} = \frac{a}{a+b}\frac{\mu Z_{lr} - \kappa Z_{lr}^2}{\mu Z_r - \kappa(Z_{rr}^2 + Z_{lr}^2)} mur \tag{5.40}$$

The combination of the bounce-pitch-roll model and the lateral-yaw model allows all eight unknown forces acting on the vehicle to be found. The nonlinearity of the tire model requires that the normal tire forces all be solved in advance.

5.2 Transient Analysis

The full vehicle model is often extended to a dynamic model, usually presented as having seven degrees of freedom. In this case, the bounce, pitch, and roll motions are augmented with suspension motions for each corner, and the yaw, lateral, and longitudinal motions are ignored. In this case, the vehicle is assumed to be vibrating around an equilibrium while travelling straight ahead at constant speed, rather than in a corner as in the steady state models. The assumption of small variables is made, such that products and squares of angular velocities are relatively less important, in order to preserve the linearity in the model. Similarly, suspension kinematic effects are also typically ignored for the sake of simplicity. The assumption of small angular motions also allows the direct integration of angular velocities to find angular motions, while in fully nonlinear models this is not possible. For a more complete discussion on measures of angular orientation, see Chapter 6.

The vertical equation of motion is written as before, but equilibrium is no longer assumed. The motions are now assumed to be measured from the equilibrium position, so that the effects of gravity can be ignored. (The weight is exactly balanced by the preload forces in the suspension, and cancels out.)

$$\sum Z = -Z_{\text{rf}} - Z_{\text{lf}} - Z_{\text{rr}} - Z_{\text{lr}} = m\ddot{z} \tag{5.41}$$

The roll moment equation is also rewritten to include roll angular acceleration.

$$\sum L = \frac{t_{\text{f}}}{2}(Z_{\text{lf}} - Z_{\text{rf}}) + \frac{t_{\text{r}}}{2}(Z_{\text{lr}} - Z_{\text{rr}}) = I_{\text{xx}}\dot{p} = I_{\text{xx}}\ddot{\phi} \tag{5.42}$$

Similarly, the pitch moment equation includes pitch angular acceleration.

$$\sum M = a(Z_{\text{rf}} + Z_{\text{lf}}) - (bZ_{\text{rr}} + Z_{\text{lr}}) = I_{\text{yy}}\dot{q} = I_{\text{yy}}\ddot{\theta} \tag{5.43}$$

Each of suspension forces can be expressed as function of the compression of each main spring, the twist of the anti-roll bars, and the compression rate of the dampers.

$$\begin{aligned} Z_{\text{rf}} = {} & k_{\text{f}}\left(z - a\theta + \frac{1}{2}t_{\text{f}}\phi - z_{\text{rf}}\right) + c_{\text{f}}\left(\dot{z} - a\dot{\theta} + \frac{1}{2}t_{\text{f}}\dot{\phi} - \dot{z}_{\text{rf}}\right) \\ & + k_{\text{fb}}(z_{\text{lf}} - z_{\text{rf}} + t_{\text{f}}\phi) \end{aligned} \tag{5.44}$$

$$\begin{aligned} Z_{\text{lf}} = {} & k_{\text{f}}\left(z - a\theta - \frac{1}{2}t_{\text{f}}\phi - z_{\text{lf}}\right) + c_{\text{f}}\left(\dot{z} - a\dot{\theta} - \frac{1}{2}t_{\text{f}}\dot{\phi} - \dot{z}_{\text{lf}}\right) \\ & + k_{\text{fb}}(z_{\text{rf}} - z_{\text{lf}} - t_{\text{f}}\phi) \end{aligned} \tag{5.45}$$

$$\begin{aligned} Z_{\text{rr}} = {} & k_{\text{r}}\left(z + b\theta + \frac{1}{2}t_{\text{r}}\phi - z_{\text{rr}}\right) + c_{\text{r}}\left(\dot{z} + b\dot{\theta} + \frac{1}{2}t_{\text{r}}\dot{\phi} - \dot{z}_{\text{rr}}\right) \\ & + k_{\text{rb}}(z_{\text{lr}} - z_{\text{rr}} + t_{\text{r}}\phi) \end{aligned} \tag{5.46}$$

$$\begin{aligned} Z_{\text{lr}} = {} & k_{\text{r}}\left(z + b\theta - \frac{1}{2}t_{\text{r}}\phi - z_{\text{lr}}\right) + c_{\text{r}}\left(\dot{z} + b\dot{\theta} - \frac{1}{2}t_{\text{r}}\dot{\phi} - \dot{z}_{\text{lr}}\right) \\ & + k_{\text{rb}}(z_{\text{rr}} - z_{\text{lr}} - t_{\text{r}}\phi) \end{aligned} \tag{5.47}$$

Each of the four unsprung masses also has an equation of motion.

$$m_{\text{f}}\ddot{z}_{\text{rf}} = -k_{\text{ft}}(z_{\text{rf}} - u_{\text{rf}}) + Z_{\text{rf}} \tag{5.48}$$

$$m_{\text{f}}\ddot{z}_{\text{lf}} = -k_{\text{ft}}(z_{\text{lf}} - u_{\text{lf}}) + Z_{\text{lf}} \tag{5.49}$$

$$m_r \ddot{z}_{rr} = -k_{rt}(z_{rr} - u_{rr}) + Z_{rr} \tag{5.50}$$

$$m_r \ddot{z}_{lr} = -k_{rt}(z_{lr} - u_{lr}) + Z_{lr} \tag{5.51}$$

The combined equations of motion have the form:

$$\mathbf{M}\ddot{\boldsymbol{z}} + \mathbf{L}\dot{\boldsymbol{z}} + \mathbf{K}\boldsymbol{z} = \mathbf{F}\boldsymbol{u} \tag{5.52}$$

where:

$$\boldsymbol{z} = \begin{bmatrix} z & \phi & \theta & z_{rf} & z_{lf} & z_{rr} & z_{lr} \end{bmatrix}' \tag{5.53}$$

and:

$$\mathbf{M} = \begin{bmatrix} m & 0 & 0 & 0 & 0 & 0 & 0 \\ 0 & I_{xx} & 0 & 0 & 0 & 0 & 0 \\ 0 & 0 & I_{yy} & 0 & 0 & 0 & 0 \\ 0 & 0 & 0 & m_f & 0 & 0 & 0 \\ 0 & 0 & 0 & 0 & m_f & 0 & 0 \\ 0 & 0 & 0 & 0 & 0 & m_r & 0 \\ 0 & 0 & 0 & 0 & 0 & 0 & m_r \end{bmatrix} \tag{5.54}$$

$$\mathbf{L} = \begin{bmatrix} 2(c_f + c_r) & 0 & 2(bc_r - ac_f) & -c_f & -c_f & -c_r & -c_r \\ 0 & \frac{1}{2}(t_f^2 c_f + t_r^2 c_r) & 0 & -\frac{1}{2}t_f c_f & \frac{1}{2}t_f c_f & -\frac{1}{2}t_r c_r & \frac{1}{2}t_r c_r \\ 2(bc_r - ac_f) & 0 & 2(a^2 c_f + b^2 c_r) & ac_f & ac_f & -bc_r & -bc_r \\ -c_f & -\frac{1}{2}t_f c_f & ac_f & c_f & 0 & 0 & 0 \\ -c_f & \frac{1}{2}t_f c_f & ac_f & 0 & c_f & 0 & 0 \\ -c_r & -\frac{1}{2}t_r c_r & -bc_r & 0 & 0 & c_r & 0 \\ -c_r & \frac{1}{2}t_r c_r & -bc_r & 0 & 0 & 0 & c_r \end{bmatrix} \tag{5.55}$$

and:

$$\mathbf{K} = \begin{bmatrix} 2(k_f + k_r) & 0 & 2(bk_r - ak_f) & -k_f & -k_f & -k_r & -k_r \\ 0 & \frac{1}{2}(t_f^2 k_{fe} + t_r^2 k_{re}) & 0 & -\frac{1}{2}t_f k_{fe} & \frac{1}{2}t_f k_{fe} & -\frac{1}{2}t_r k_{re} & \frac{1}{2}t_r k_{re} \\ 2(bk_r - ak_f) & 0 & 2(a^2 k_f + b^2 k_r) & ak_f & ak_f & -bk_r & -bk_r \\ -k_f & -\frac{1}{2}t_f k_{fe} & ak_f & k_{fs} & -k_{fb} & 0 & 0 \\ -k_f & \frac{1}{2}t_f k_{fe} & ak_f & -k_{fb} & k_{fs} & 0 & 0 \\ -k_r & -\frac{1}{2}t_r k_{re} & -bk_r & 0 & 0 & k_{rs} & -k_{rb} \\ -k_r & \frac{1}{2}t_r k_{re} & -bk_r & 0 & 0 & -k_{rb} & k_{rs} \end{bmatrix} \tag{5.56}$$

where:

$$k_{fs} = k_f + k_{fb} + k_{ft} \tag{5.57}$$

$$k_{rs} = k_r + k_{rb} + k_{rt} \tag{5.58}$$

Note that the damping matrix is exactly the same form as the stiffness matrix, with the anti-roll and tire stiffnesses set to zero, and the main spring stiffnesses replaced with the appropriate damping coefficient. The input matrix is simply:

$$\mathbf{F} = \begin{bmatrix} 0 & 0 & 0 & 0 \\ 0 & 0 & 0 & 0 \\ 0 & 0 & 0 & 0 \\ k_{ft} & 0 & 0 & 0 \\ 0 & k_{ft} & 0 & 0 \\ 0 & 0 & k_{rt} & 0 \\ 0 & 0 & 0 & k_{rt} \end{bmatrix} \tag{5.59}$$

The transient analysis of the full car model can proceed in a standard fashion. The natural frequencies of motion can be found by taking the square roots of the eigenvalues of the $\mathbf{M}^{-1}\mathbf{K}$ matrix, or by recasting the equations into first order form, which is required if damping effects are included. The problem is now much too big to be solved by hand, so a numerical solution must be used. The results of sample calculation using the values in Table 5.1 is shown in Table 5.2. The full car model produces seven natural frequencies. Three are low frequency body motions $\approx$ 1 Hz, consisting of a pair of coupled bounce and pitch motions, and a roll motion. Due to the left-right symmetry of the model, the roll motion is decoupled from the bounce and pitch. The four remaining frequencies will be wheel hop modes, $\approx$ 10 Hz. The four wheel hop motions will not be hop of

Table 5.1 Full car model parameter values

Quantity	Notation	Value
front axle dimension	a	1.189 m
rear axle dimension	b	1.696 m
sprung mass	m	1 730 kg
roll inertia	I_{xx}	818 kg m^2
pitch inertia	I_{yy}	3 267 kg m^2
front suspension stiffness	k_f	17 500 N/m
front anti-roll stiffness	k_{fb}	3 000 N/m
rear suspension stiffness	k_r	19 000 N/m
rear anti-roll stiffness	k_{fb}	1 000 N/m
front suspension damping	c_f	1 000 Ns/m
rear suspension damping	c_r	1 200 Ns/m
front tire stiffness	k_{ft}	180 000 N/m
rear tire stiffness	k_{rt}	180 000 N/m
front track width	t_f	1.595 m
rear track width	t_r	1.631 m
front unsprung mass	m_f	35 kg
rear unsprung mass	m_r	30 kg

Table 5.2 Full car model vibration metrics

No.	ω_n [Hz]	ζ	τ [s]	λ [s]
1	1.1907×10^{1}	1.9316×10^{-1}	6.9198×10^{-2}	8.5597×10^{-2}
2	1.2104×10^{1}	1.9017×10^{-1}	6.9142×10^{-2}	8.4151×10^{-2}
3	1.3100×10^{1}	2.4534×10^{-1}	4.9518×10^{-2}	7.8740×10^{-2}
4	1.2851×10^{1}	2.5137×10^{-1}	4.9268×10^{-2}	8.0393×10^{-2}
5	1.3066×10^{0}	1.6341×10^{-1}	7.4545×10^{-1}	7.7579×10^{-1}
6	1.1355×10^{0}	2.0301×10^{-1}	6.9046×10^{-1}	8.9944×10^{-1}
7	9.0512×10^{-1}	1.4786×10^{-1}	1.1892×10^{0}	1.1171×10^{0}

Note: table includes natural frequencies (ω_n), damping ratios (ζ), time constants (τ), and wavelengths (λ)

the individual wheels, but rather a front wheel hop with the wheels moving in phase, a front wheel hop with the wheels moving out of phase, a rear wheel hop with the wheels moving in phase, and a rear wheel hop with the wheels moving out of phase.

5.3 Kinematic Effects

One of the shortcomings of the traditional full car model is that it ignores the effect of suspension kinematics. In reality, the suspension motion is not constrained to a vertical axis relative to the vehicle, but rather one that is both angled and curved. Typically, the track width will increase slightly as the vehicle squats down on the suspension. The result is a coupling between the lateral and vertical dynamics.

The suspension of a road vehicle is a spatial mechanism with nonlinear relationships governing the relative locations of the various components. These nonlinearities pose two possible complications to the modeller. First, if one is analyzing the motion of a vehicle in a situation where there is significant suspension motion, e.g., when predicting the loads in the individual components when travelling on rough terrain, then it may be important to include the nonlinear kinematics in the dynamic model. However, in many cases, these relations can be linearized to allow a linear dynamic model without a significant loss in fidelity. The other issue is that the kinematic properties themselves may influence the overall vehicle motion.

There are number of properties of interest. These include the camber and steer angles, and the lateral scrub of the tire as these influence the lateral force generated by the tire. Also important are changes to the *caster* and *kingpin* angles of the steer axis, as they influence the stability and driver feedback in the steering system. One of the more interesting aspects are the instant centres of roll motion, more commonly referred to as the *roll centres*.

5.3.1 Roll Centres

The roll centre is an important property of the suspension; its location is an indicator of the manner in which the tire forces must pass through the suspension into the vehicle body. The roll centre is defined as the axis around which chassis roll can occur without any corresponding lateral translation of either tire contact patch. The location is a function of the geometry of the suspension alone, and its effect is a coupling of the roll and vertical motions of the chassis. For a normal symmetric passenger car, at equilibrium, the roll centre will fall on the vehicle centreline, typically between the ground and the centre of mass. An example for an A-arm style suspension is shown in Figure 5.2. It is important to note that in general, the roll centre is not in a fixed location relative to

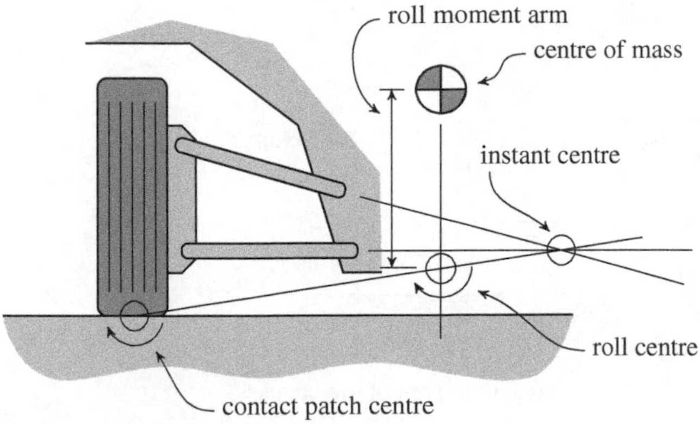

Figure 5.2 The roll centre for the A-arm type suspension can be located using a geometric approach. The A-arm geometry determines the location of the instant centre, which in turn can be used to locate the roll centre. If the vehicle chassis rotates about the roll centre, the suspension does not require any lateral scrub of the tire, but rather rotates the tire around the tire contact point. The diagram assumes that the mechanism is planar, but the same concept applies even if the geometry is more complex.

the ground, or relative to the vehicle. If the vehicle bounces, the roll centre will shift vertically; if the vehicle rolls, the roll centre will shift laterally. The distance between the roll centre and the centre of mass is the roll moment arm; this distance influences the amount of roll the vehicle will experience when cornering. As the height of the roll centre is increased, the roll moment arm is reduced, and as a result, the effective roll stiffness is increased.

A frequent question about the roll centre is the effect if it is placed at the same height as the centre of mass, to reduce the roll moment arm length to zero, thereby eliminating body roll in cornering. Consider the forces acting at a front tire contact point, as shown in Figure 5.3. Taking m_{f} as the fraction of the vehicle mass that is supported by the front axle, and assuming the roll stiffness is distributed approximately in proportion to the weight, the steady state change in normal force can be computed from the lateral acceleration as:

$$\Delta Z_{\mathrm{f}} = \frac{m_{\mathrm{f}} u r h_{\mathrm{G}}}{t} \tag{5.60}$$

Similarly, ignoring the weight transfer effect on the tire's cornering stiffness, and assuming the slip angles are equal on the left and right, the lateral force at a single front tire can be found.

$$Y_{\mathrm{f}} = \frac{m_{\mathrm{f}} u r}{2} \tag{5.61}$$

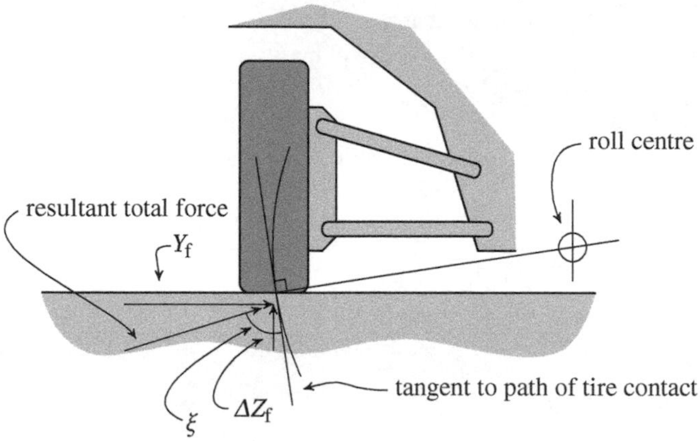

Figure 5.3 If the roll centre is quite high, the direction of the applied cornering force will cause the suspension to go into rebound rather than compression.

The tangent of the angle of the resultant applied force is then:

$$\frac{\Delta Z_f}{Y_f} = \frac{\frac{m_f u r h_G}{t}}{\frac{m_f u r}{2}} = \frac{h_G}{t/2} \tag{5.62}$$

From the geometry of the tangent to the path of the contact point, an expression for the height of the roll centre can be found.

$$\frac{\Delta y}{\Delta z} = \frac{h_{RC}}{t/2} \tag{5.63}$$

If the applied force acts perpendicular to the tangent to the path of the tire contact point, then no compression or rebound of the suspension occurs in response to the applied force, and as a result, no roll motion occurs under cornering. The effect can be quantified by the angle ξ, between the resultant force and the tangent, as shown in Figure 5.3.

$$\xi = \frac{\pi}{2} + \frac{h_{RC} - h_G}{t/2} \tag{5.64}$$

Clearly, if the roll centre is placed at the same height of the centre of mass, then $\xi = \frac{\pi}{2}$. However, while theoretically possible, in most cases, it is highly impractical to do so. High roll centres are usually associated with large amounts of lateral tire scrub during suspension travel. In order to place the roll centre on the centre of mass, the suspension would often require the wheel to move laterally as much or more than its vertical motion, which would have a number of detrimental effects on ride and handling. If, in this case, a single wheel struck a bump, its lateral motion would generate a slip angle and corresponding lateral force. The vehicle would have lateral response to vertical road disturbances.

Another factor that should be considered when selecting roll centre locations is the effect of *jacking*. If the roll centre is high enough, when subjected to lateral tire forces, the suspension will react by lifting the vehicle, allowing the wheel to 'tuck under' the body. Lateral forces acting on the tire are effectively reacted on the body at the height of the roll centre, generating roll moments in the suspension mechanism that must be balanced by additional vertical forces, called *jacking forces*. The total jacking force contribution would be zero, if both the outer and inner tire contributed the same amount of lateral force. However, this is not the case. The increase in normal force on the outer tire due to weight transfer will cause its lateral force contribution to be larger than that of the inner tire. As a result, its jacking force will also be larger. The lateral force on the outer tire will generate a lifting force, while the lateral force on the inner tire will generate a squatting force. The two forces will not be equal and opposite, with the larger outer force resulting in a net lifting force. The effect will be a lifting of the chassis as the vehicle corners, if the roll centre is above the ground, as is the usual case. The effect is illustrated in a free body diagram in Figure 5.4.

Generally, as the body lifts in response to jacking forces, the roll centre height increases, amplifying the effect. To further complicate matters, large suspension extensions are often accompanied by positive tire camber, which tends to reduce grip. In the worst case, under hard cornering, a high rear roll centre can result in a dramatic, even violent transition to severe oversteer behaviour, with its associated yaw instability. In most modern passenger vehicles, the roll centre is low enough that the jacking effect is very mild. However, for a number of years, before auto manufacturers fully understood the implications of the roll centre location, the use of swing axle style rear suspension was quite popular. The roll centre of swing axle style suspension is notably high. Today, these swing axle vehicles are remembered for being notorious for their particularly challenging handling characteristics.

5.3.2 Quasi Static Model, with Roll Centres

The effect of the suspension kinematics on the full car model can be included by modifying the equations of motion slightly. The vertical dynamics are influenced by the effect of jacking forces. These are the resulting forces that are acting in the suspension, caused by the geometry of the individual arms. In order for the mechanism to be in equilibrium when lateral forces are acting on the tire, a resulting vertical force is required for the arms to achieve a moment balance.

In order to include jacking effects in the model, a simple suspension with equivalent kinematics around the equilibrium point is added. The suspension is two swing arms, pinned to the chassis at the roll centre. A moment balance for each swing arm provides four extra equations. The arms are assumed to

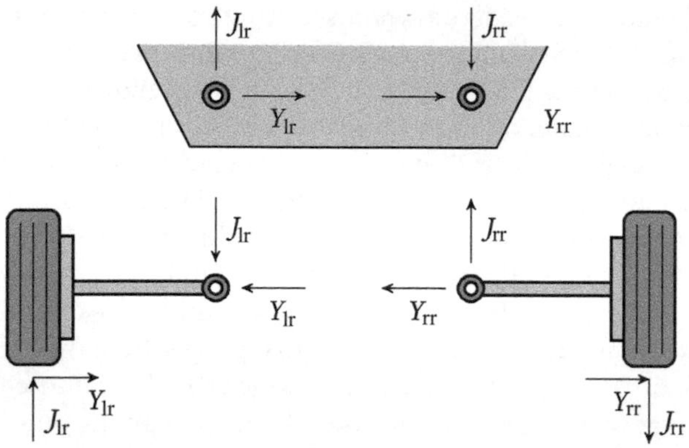

Figure 5.4 The source of jacking forces is apparent when considering a free body diagram of an individual swing axle. In this case, the force Y_{lr} would typically be larger than Y_{rr}, and as a result, J_{lr} would be larger than J_{rr}, giving a net lifting force. The spring forces are neglected in this figure as they have only a small influence on the jacking behaviour.

be in steady state, and their mass is ignored, as it is significantly less than the sprung mass.

$$\frac{t_f}{2}J_{rf} - h_f Y_{rf} = 0 \tag{5.65}$$

$$\frac{t_f}{2}J_{lf} - h_f Y_{lf} = 0 \tag{5.66}$$

$$\frac{t_r}{2}J_{rr} - h_r Y_{rr} = 0 \tag{5.67}$$

$$\frac{t_r}{2}J_{lr} - h_r Y_{lr} = 0 \tag{5.68}$$

The equations of motion for the chassis are now modified to include the jacking forces.

$$\sum Z = -Z_{rf} - Z_{lf} - Z_{rr} - Z_{lr} + J_{rf} - J_{lf} + J_{rr} - J_{lr} + mg = m\ddot{z} = 0 \tag{5.69}$$

Before the roll moment equation is prepared, a new reference point must be selected. If the reference is taken at the ground, as in the first case, the lateral forces applied by the suspension would contribute to the roll moment. If the rear lateral force acts at height h_r, and the front lateral force acts at height h_f, then the roll moment generated by these two can be equated to a moment around an equivalent point where all four lateral forces act. The effect is illustrated in Figure 5.5.

$$\begin{aligned}\sum L &= (h_G - h_f)(Y_{rf} + Y_{lf}) + (h_G - h_r)(Y_{rr} + Y_{lr}) \\ &= (h_G - h_{RC})(Y_{rf} + Y_{lf} + Y_{rr} + Y_{lr})\end{aligned} \tag{5.70}$$

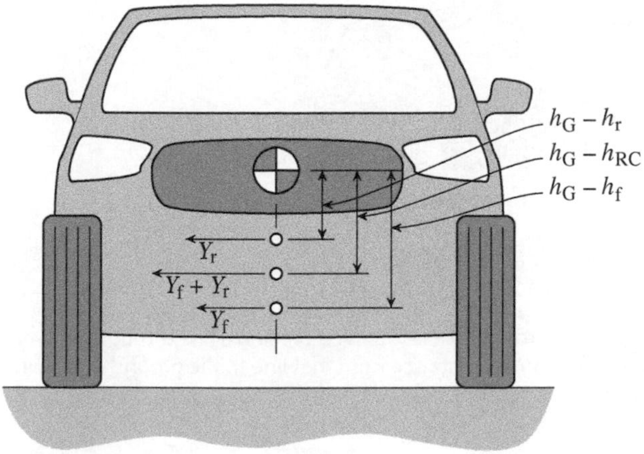

Figure 5.5 When including roll centre effects, the free body diagram of the chassis will show lateral forces acting at the roll centres.

Now, if the vehicle is in yaw equilibrium, then the yaw equation is:

$$\sum N = a(Y_{rf} + Y_{lf}) - b(Y_{rr} + Y_{lr}) = I_{zz}\dot{r} = 0 \tag{5.71}$$

or:

$$\frac{b}{a} = \frac{Y_{rf} + Y_{lf}}{Y_{rr} + Y_{lr}} \tag{5.72}$$

Dividing Equation (5.70) by $Y_{rr} + Y_{lr}$ and substituting Equation (5.72) into the result gives:

$$\left(\frac{b}{a} + 1\right)(h_G - h_{RC}) = \frac{b}{a}(h_G - h_f) + (h_G - h_r) \tag{5.73}$$

Or, rearranging:

$$(h_G - h_{RC}) = \frac{a(h_G - h_r) + b(h_G - h_f)}{a + b} \tag{5.74}$$

A careful examination of the result shows that the effective point where all four of the lateral forces act is at the intersection of a line connecting the front and rear roll centres, and a line extended vertically below the centre of mass. The effect is illustrated in Figure 5.6. The roll moment equation is modified to sum moments around a longitudinal axis that intersects this roll centre point. The expressions show that a sum of the moments around this point is equivalent to:

$$\sum L = \frac{t_f}{2}(Z_{rf} - Z_{lf}) + \frac{t_r}{2}(Z_{rr} - Z_{lr}) = mur(h_G - h_{RC}) \tag{5.75}$$

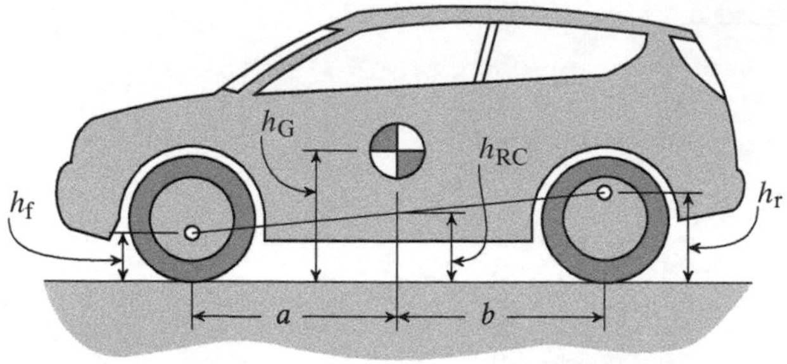

Figure 5.6 The effective height of the roll centre is found by passing a line through each of the front at rear roll centres, and finding the distance from that line to the ground, at a point directly beneath the centre of mass.

Finally, the pitch moment equation can be included, also modified to include the jacking forces.

$$\sum M = a(Z_{rf} + Z_{lf} + J_{rf} - J_{lf}) - b(Z_{rr} + Z_{lr} + J_{rr} - J_{lr})$$
$$= I_{yy}\dot{q} - I_{xz}r^2 = -I_{xz}r^2 \tag{5.76}$$

The equations can now be assembled.

$$\begin{bmatrix} 1 & 1 & 1 & 1 & -2\frac{h_f}{t_f} & 2\frac{h_f}{t_f} & -2\frac{h_r}{t_r} & 2\frac{h_r}{t_r} & 0 & 0 & 0 \\ -t_f & t_f & -t_r & t_r & 0 & 0 & 0 & 0 & 0 & 0 & 0 \\ a & a & -b & -b & 2a\frac{h_f}{t_f} & -2a\frac{h_f}{t_f} & 2b\frac{h_r}{t_r} & -2b\frac{h_r}{t_r} & 0 & 0 & 0 \\ 0 & 0 & 0 & 0 & 1 & 1 & 1 & 1 & 0 & 0 & 0 \\ 0 & 0 & 0 & 0 & a & a & -b & -b & 0 & 0 & 0 \\ -1 & 0 & 0 & 0 & 0 & 0 & 0 & 0 & k_f & \frac{1}{2}t_f k_{fe} & -ak_f \\ 0 & -1 & 0 & 0 & 0 & 0 & 0 & 0 & k_f & -\frac{1}{2}t_f k_{fe} & -ak_f \\ 0 & 0 & -1 & 0 & 0 & 0 & 0 & 0 & k_r & \frac{1}{2}t_r k_{re} & bk_r \\ 0 & 0 & 0 & -1 & 0 & 0 & 0 & 0 & k_r & -\frac{1}{2}t_r k_{re} & bk_r \end{bmatrix} \begin{Bmatrix} Z_{rf} \\ Z_{lf} \\ Z_{rr} \\ Z_{lr} \\ Y_{rf} \\ Y_{lf} \\ Y_{rr} \\ Y_{lr} \\ z \\ \phi \\ \theta \end{Bmatrix}$$

$$= \left[mg \;\; 2mur(h_G - h_{RC}) \;\; -I_{xz}r^2 \;\; mur \;\; 0 \;\; 0 \;\; 0 \;\; 0 \;\; 0\right]' \tag{5.77}$$

Of course, in the form given, there are eleven unknowns, but only nine equations. In order to solve, the set of equations would need to be supplemented with two additional equations describing the tire properties. For example, consider the example tire model:

$$Y_{\mathrm{rf}} = f(\alpha_{\mathrm{rf}})(\mu(Z_{\mathrm{rf}} - J_{\mathrm{rf}}) - \kappa(Z_{\mathrm{rf}} - J_{\mathrm{rf}})^2) \tag{5.78}$$

where μ and κ are constants. In this case:

$$f(\alpha_{\mathrm{rf}}) = \frac{Y_{\mathrm{rf}}}{\mu(Z_{\mathrm{rf}} - J_{\mathrm{rf}}) - \kappa(Z_{\mathrm{rf}} - J_{\mathrm{rf}})^2} \tag{5.79}$$

If the front left and front right tire are assumed to have the same slip angle, i.e., if $\alpha_{\mathrm{rf}} = \alpha_{\mathrm{lf}}$, then $f(\alpha_{\mathrm{rf}}) = f(\alpha_{\mathrm{lf}})$, or:

$$\frac{Y_{\mathrm{rf}}}{\mu(Z_{\mathrm{rf}} - J_{\mathrm{rf}}) - \kappa(Z_{\mathrm{rf}} - J_{\mathrm{rf}})^2} = \frac{Y_{\mathrm{lf}}}{\mu(Z_{\mathrm{lf}} + J_{\mathrm{lf}}) - \kappa(Z_{\mathrm{lf}} + J_{\mathrm{lf}})^2} \tag{5.80}$$

and by symmetry:

$$\frac{Y_{\mathrm{rr}}}{\mu(Z_{\mathrm{rr}} - J_{\mathrm{rr}}) - \kappa(Z_{\mathrm{rr}} - J_{\mathrm{rr}})^2} = \frac{Y_{\mathrm{lr}}}{\mu(Z_{\mathrm{lr}} + J_{\mathrm{lr}}) - \kappa(Z_{\mathrm{lr}} + J_{\mathrm{lr}})^2} \tag{5.81}$$

Here, the expressions representing the tire behaviour are clearly nonlinear, and cannot be combined directly with the linear set above. In order to solve them along with the linear set, an iterative scheme would be required. For example, two equations for estimates of Y_{rf} and Y_{rr} could be appended to the linear set, and the remaining nine unknowns could be solved. Those nine values can then be substituted in the tire models to get updated values of the estimates. The improved estimates are then used to repeat the solution of the nine unknowns. The process continues, until the values stabilize. Of course, this process requires that that solution will converge, and this is not guaranteed. The less severe the nonlinearity of the tire model, the greater the likelihood of convergence. In any case, once the roll centre effects are included in the model, the vertical and lateral force problems are no longer decoupled; they must be solved simultaneously, and the nonlinearity of the tire model will require that an iterative process is used.

5.4 Numerical Solution of Suspension Kinematics

Often, as a part of the suspension design process, it becomes desirable to solve the suspension kinematics problem separately from the dynamics. There are two mathematical approaches that can be used to do so. These will be described here with the double A-arm type suspension used as an example. The first approach proceeds by formulating a series of equations that describe the required relationships between the locations of the various components, often referred to as *constraint equations.*

5.4.1 Algebraic Equations Formulation

In this case, in mathematical terms, the kinematics problem is a system of coupled nonlinear algebraic equations. If each constraint is a function of the coordinates $\boldsymbol{p}$, then they can be collected in a vector as:

$$\boldsymbol{\phi}(\boldsymbol{p}) = \mathbf{0} \tag{5.82}$$

Typically, the number of coordinates (the length of the vector $\boldsymbol{p}$) would be larger than the number of constraints (the length of the vector $\boldsymbol{\phi}$) by one, with the difference corresponding to the one degree of freedom associated with the suspension travel. A typical double A-arm suspension, as shown in Figure 5.7, can be modelled with five constraints, chosen as follows: pick any two points that lie on the axis of rotation of the upper A-arm, e.g., the mounting points of the arm on the chassis. The distance from each of these two points to the upper ball joint must be a constant value, regardless of the motion of the upright. The same process can be repeated for the lower A-arm, and lower ball joint. Finally, the distance between the inner and out tie-rod ends must also be fixed. The equations can be written:

$$\boldsymbol{\phi} = \begin{Bmatrix} (x_{\mathrm{A}} - x_{\mathrm{UBJ}})^2 + (y_{\mathrm{A}} - y_{\mathrm{UBJ}})^2 + (z_{\mathrm{A}} - z_{\mathrm{UBJ}})^2 - l^2_{\mathrm{A\text{-}UBJ}} \\ (x_{\mathrm{B}} - x_{\mathrm{UBJ}})^2 + (y_{\mathrm{A}} - y_{\mathrm{UBJ}})^2 + (z_{\mathrm{A}} - z_{\mathrm{UBJ}})^2 - l^2_{\mathrm{B\text{-}UBJ}} \\ (x_{\mathrm{C}} - x_{\mathrm{LBJ}})^2 + (y_{\mathrm{A}} - y_{\mathrm{LBJ}})^2 + (z_{\mathrm{A}} - z_{\mathrm{LBJ}})^2 - l^2_{\mathrm{C\text{-}LBJ}} \\ (x_{\mathrm{D}} - x_{\mathrm{LBJ}})^2 + (y_{\mathrm{A}} - y_{\mathrm{LBJ}})^2 + (z_{\mathrm{A}} - z_{\mathrm{LBJ}})^2 - l^2_{\mathrm{D\text{-}LBJ}} \\ (x_{\mathrm{E}} - x_{\mathrm{TRE}})^2 + (y_{\mathrm{A}} - y_{\mathrm{TRE}})^2 + (z_{\mathrm{A}} - z_{\mathrm{TRE}})^2 - l^2_{\mathrm{E\text{-}TRE}} \end{Bmatrix} = \begin{Bmatrix} 0 \\ 0 \\ 0 \\ 0 \\ 0 \end{Bmatrix} \tag{5.83}$$

However, in this case, there is a problem. There are six motions possible of the upright, but there are nine coordinates used to specify the location of the three points fixed to the upright. Three additional constraints are required to fix the

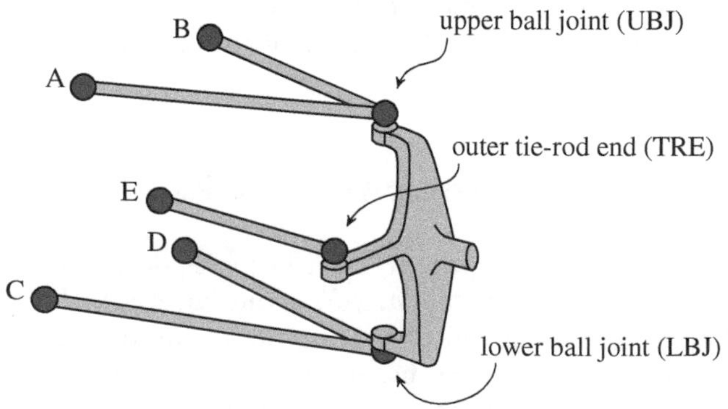

Figure 5.7 The kinematics of the A-arm suspension can be solved for camber angle, toe angle, scrub, etc., as a function of vertical displacement.

relative locations of the three points; these can be appended to the existing list.

$$\left\{\begin{matrix} (x_{\mathrm{UBJ}} - x_{\mathrm{LBJ}})^2 + (y_{\mathrm{UBJ}} - y_{\mathrm{LBJ}})^2 + (z_{\mathrm{UBJ}} - z_{\mathrm{LBJ}})^2 - l^2_{\mathrm{UBJ\text{-}LBJ}} \\ (x_{\mathrm{TRE}} - x_{\mathrm{LBJ}})^2 + (y_{\mathrm{TRE}} - y_{\mathrm{LBJ}})^2 + (z_{\mathrm{TRE}} - z_{\mathrm{LBJ}})^2 - l^2_{\mathrm{TRE\text{-}LBJ}} \\ (x_{\mathrm{TRE}} - x_{\mathrm{UBJ}})^2 + (y_{\mathrm{TRE}} - y_{\mathrm{UBJ}})^2 + (z_{\mathrm{TRE}} - z_{\mathrm{UBJ}})^2 - l^2_{\mathrm{TRE\text{-}UBJ}} \end{matrix}\right\} = \left\{\begin{matrix} 0 \\ 0 \\ 0 \end{matrix}\right\} \tag{5.84}$$

With the three additional equations there are now nine unknowns, and eight equations. A value for any one of the nine unknowns can be chosen (e.g., the z coordinate of the lower ball joint), and the remaining eight unknowns can be found by solving the set of equations. In this case, the specific value chosen for that unknown might be varied over a range of values to find the solutions corresponding to that range.

This set of equations can be solved using an iterative process such as the Newton–Raphson algorithm, where an initial guess at the solution is repeatedly refined based on the error in the solutions. A difficulty in dealing with systems of nonlinear algebraic equations is that there is no guarantee that a solution exists, or that the solution is unique. Even if a solution is known to exist, it is often a challenge to find an initial set of values that allows the algorithm to converge on it. For a more detailed description of the Newton–Raphson algorithm, see Chapter 7.

While the model described here is suitable for A-arm style suspensions, there are of course many other varieties (e.g., MacPherson strut, multilink, etc.), and each requires its own particular set of constraint equations. For more detail on the various mechanisms, and how to form the appropriate constraint equations, see Dixon [1].

5.4.2 Differential Equations Formulation

An alternate approach to the suspension kinematics problem is based on casting the problem as a set of differential equations. Consider the time derivative of the constraint equations, allowing that the constraints are not time-dependent, i.e., $\boldsymbol{\phi} \neq \boldsymbol{\phi}(t)$.

$$\frac{\partial \boldsymbol{\phi}}{\partial \boldsymbol{p}}\dot{\boldsymbol{p}} = \boldsymbol{0} \tag{5.85}$$

The differentiated constraints are effectively statements that the velocities of the three points on the upright must be perpendicular to the arms themselves, i.e, the rate of stretch of the arms is zero. These equations are then combined with an additional requirement that some point on the upright has a given velocity, e.g., the lower ball joint has a constant vertical speed of 1 m/s. The additional requirement is written as a linear function of the velocities:

$$\boldsymbol{\psi}(\boldsymbol{p})\dot{\boldsymbol{p}} = 1 \tag{5.86}$$

Combining the equations gives:

$$\begin{bmatrix} \dfrac{\partial \boldsymbol{\phi}}{\partial \boldsymbol{p}} \\ \boldsymbol{\psi} \end{bmatrix} \dot{\boldsymbol{p}} = \begin{Bmatrix} 0 \\ \vdots \\ 0 \\ 1 \end{Bmatrix} \tag{5.87}$$

The result is a linear system of equations in $\dot{\boldsymbol{p}}$. The solution of a linear system is much more straight forward than a nonlinear system, and there are many methods that can be used, e.g., Gaussian elimination. In practice a matrix inversion would not normally be used in the calculation as it leads to increased round-off error, but it can be used to show that $\dot{\boldsymbol{p}} = \dot{\boldsymbol{p}}(\boldsymbol{p})$, i.e., the solution is in fact an ODE, and so can be solved using a standard routine, e.g., Runge–Kutta.

$$\dot{\boldsymbol{p}} = \begin{bmatrix} \dfrac{\partial \boldsymbol{\phi}}{\partial \boldsymbol{p}} \\ \boldsymbol{\psi} \end{bmatrix}^{-1} \begin{Bmatrix} 0 \\ \vdots \\ 0 \\ 1 \end{Bmatrix} \tag{5.88}$$

One immediate advantage of this approach is that it removes the possibility of not converging on a solution. However, it does introduce the possibility of accumulating significant numerical round-off error as the solution progresses in time. Unfortunately, for some problems, one may be faced with the prospect of choosing between one method that does not provide a complete solution, or another one that produces a solution that contains some obvious inaccuracies. Another shortcoming of this approach is that an initial solution for $\boldsymbol{p}$ must be known, or found using a recursive method as described previously, to serve as the initial condition of the ODE solution.

In the case of the A-arm suspension, the problem can be recast into a slightly simpler ODE problem as follows. First, an expression relating the velocity of the upper ball joint to the velocity of the upright's mass centre is formed using simple kinematics.

$$\boldsymbol{v}_{\text{UBJ}} = \boldsymbol{v}_{\text{G}} + \boldsymbol{\omega} \times \boldsymbol{r}_{\text{UBJ/G}} \tag{5.89}$$

Next, one can recognize that if the distance between the upper ball joint and the point A is a constant, then no component of the velocity of the ball joint can lie in the direction towards A. A unit vector $\hat{\boldsymbol{u}}_{\text{A-UBJ}}$ in this direction is defined, and the velocity expression is written.

$$\boldsymbol{v}_{\text{UBJ}} \cdot \hat{\boldsymbol{u}}_{\text{A-UBJ}} = 0 \tag{5.90}$$

Using some algebraic manipulation, the velocities are factored out, and the two equations as combined and written as a row vector multiplied against a column vector.

$$\begin{bmatrix}\hat{\boldsymbol{u}}'_{\text{A-UBJ}} & (\boldsymbol{r}_{\text{UBJ/G}} \times \hat{\boldsymbol{u}}_{\text{A-UBJ}})'\end{bmatrix} \begin{Bmatrix} \boldsymbol{v}_{\text{G}} \\ \boldsymbol{\omega} \end{Bmatrix} = 0 \tag{5.91}$$

Once the first row vector is in place, similar relationship can be formed using the other suspension arms and points, adding four more rows to the matrix.

$$\begin{bmatrix}\hat{\boldsymbol{u}}'_{\text{A-UBJ}} & (\boldsymbol{r}_{\text{UBJ/G}} \times \hat{\boldsymbol{u}}_{\text{A-UBJ}})' \\ \hat{\boldsymbol{u}}'_{\text{B-UBJ}} & (\boldsymbol{r}_{\text{UBJ/G}} \times \hat{\boldsymbol{u}}_{\text{B-UBJ}})' \\ \hat{\boldsymbol{u}}'_{\text{C-LBJ}} & (\boldsymbol{r}_{\text{LBJ/G}} \times \hat{\boldsymbol{u}}_{\text{C-UBJ}})' \\ \hat{\boldsymbol{u}}'_{\text{D-LBJ}} & (\boldsymbol{r}_{\text{LBJ/G}} \times \hat{\boldsymbol{u}}_{\text{D-UBJ}})' \\ \hat{\boldsymbol{u}}'_{\text{E-TRE}} & (\boldsymbol{r}_{\text{TRE/G}} \times \hat{\boldsymbol{u}}_{\text{E-TRE}})' \\ \hat{\mathbf{k}} & \mathbf{0}\end{bmatrix} \begin{Bmatrix} \boldsymbol{v}_{\text{G}} \\ \boldsymbol{\omega} \end{Bmatrix} = \begin{Bmatrix} 0 \\ 0 \\ 0 \\ 0 \\ 0 \\ 1 \end{Bmatrix} \tag{5.92}$$

In this formulation, a linear system of equations can be solved to find the velocity and angular velocity of the upright. From these velocities, the velocities of the upper and lower ball joint, and the tire rod end can be found. In effect, the problem becomes a differential equation; the rates of change of location of the four points depend on the locations of the four points. With this arrangement, twelve velocities will need to be integrated to find the updated locations of the four points, but the system of equations to be solved at each time step is only six by six.

Finally, it should be mentioned that when dealing the with dynamics of mechanical systems, there may be additional advantages to the ODE approach that are not apparent by inspection of the general form of equations. When describing the orientation of a rigid body in three dimensions, the choice of coordinates poses certain challenges, due to the fact that no associated vector quantity exists. If the problem can be recast as an ODE, the angular positions may be replaced with angular velocities, which are a vector quantity, and can allow significant simplification of the problem.

5.4.3 Tire Orientation Effects

Using either of the methods above allows a full kinematic analysis to be conducted, so that the motion and orientation of the wheel relative to the chassis can be found. Of all the kinematic data, the properties of camber angle, steer angle, and scrub are of significant interest, because they influence the tire force. Because they are so influential on the vehicle motion, it is key that the suspension limits the variation of these quantities. The steer and camber angle of the tire are directly related to the tire force, but the relationship to scrub is somewhat less obvious. However, if one considers that the vehicle is in forward motion while the suspension motion occurs, then the resulting scrub is effectively contributing to the tire slip angle, by generating a lateral velocity at the tire. In the event of a one-wheel bump, i.e., a case where one side of the road is smooth and the other is rough, a vehicle fitted with a suspension that has significant scrub will be disturbed laterally by the resulting imbalanced tire forces. Of course, the scrub is also related to the roll centre height as described in Equation (5.63).

Bump Steer

The terms *bump steer* or *roll steer* are used to describe the coupling between suspension motion and steering angle. Before discussing bump steer, it is important to define *toe* as the difference in the steer angle between the left side and right side tires. As one might anticipate, the term *toe-in* refers to the case where both tires are steered slightly inward toward the chassis; *toe-out* is the opposite case, where both tires are steered slightly outward. Most manufacturers specify a very small amount of static toe-in on the front axle of rear wheel drive vehicles, as it tends to improve straight line tracking at highway speeds. On front wheel drive vehicles, there is often a small amount of toe-in that results from elastic deflection of the suspension mounts due to the traction forces, so the static toe-in values are often reduced. Independently of the static values of toe, there will be additional steer that occurs during the suspension travel as a result of bump steer. Similar to scrub, this effect will also tend to disturb the vehicle laterally in the event of one-wheel bump.

There two distinct cases of steer and suspension coupling, both based on the geometry of the steering and suspension systems. In the first case, if the suspension is deflected, the direction of the resulting change in steer angle is independent of whether the suspension motion is in compression or rebound. This occurs when the steering tie-rods are of the incorrect length. When the suspension travels, the outer tie-rod end and the upper and lower ball-joints travel on arced paths, as shown in Figure 5.8. If the tie-rod is too long, the arc of

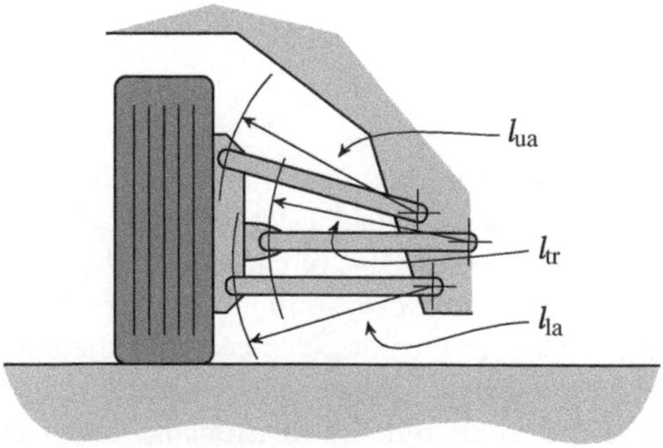

Figure 5.8 The bump steer effect is caused by the difference in the path that each of the ball-joints and tie-rod ends must follow.

the outer tie-rod end will have a larger radius than those of the upper and lower ball-joints, and cause the tie-rod end to move laterally relative to the upper and lower ball-joints, which in turn forces the upright to steer. If the tie-rods are located behind the front axle, as is usually the case, this motion will result in toe-in; conversely, if the tie-rods are too short, the radius will be too small, and toe-out will result. Vehicles where the tie-rods lie in front of the axle will have the reverse relationship. It should also be noted that wheel alignment procedures done during routine vehicle maintenance may entail small adjustments in the length of the tie-rods, to correct for errors in the static values of toe-in or toe-out. These adjustments are small enough not to have a significant effect on the path of the tie-rod end, and are not considered to influence bump steer.

In the second case, the direction of the resulting steer angle will reverse if the direction of the suspension motion is reversed. This is typically caused by an incorrect vertical location of the steering rack. If the location of the rack is vertically above the ideal location, then suspension compression will cause toe-in, and rebound will cause toe-out, again, assuming the tie-rods lie behind the front axle. If the rack is mounted vertically below the ideal location, compression will cause toe-out, and rebound will cause toe-in. In this case, the coupling between the roll and steer motions is significant. If the vehicle is in a corner, the resulting compression of the outer suspensions will generate a steer angle influence, e.g, toe-in. On the inner corner, the suspension will be in rebound, and will generate the opposite effect, toe-out. This combination effectively increases the total steer angle. Too much roll steer effect can impair the driver's ability to precisely control the heading of the vehicle, as the transient roll angle changes the driver's intended steer input.

In most vehicles, there will be some bump steer effect, as it is practically impossible to eliminate entirely, but generally the design should keep it to a minimum. In some cases, roll steer effects may be intentionally engineered into the rear axle, in order to generate some small steer angle in response to body roll, in an attempt to influence handling.

Ackermann Steering

It is also important to note that most vehicles are intentionally designed to toe-out as the steer angle is increased, a feature referred to as *Ackermann steering*, and independent of the amount of bump steer present. The logic behind Ackermann steering is that the inner wheel is following a curved path on the road of smaller radius than the outer wheel, and so it should be steered more to compensate. The desired relationship between the steer angle of the outer wheel δ_o and the steer angle of the inner wheel δ_i is given by:

$$\cot \delta_o - \cot \delta_i = \frac{t}{a+b} \tag{5.93}$$

There are two relevant points here. First, the geometry of the rack and pinion steering mechanism that is typically used to generate the Ackermann effect can be analyzed to find the actual relationship between the inner and outer steer angles. If this relationship is compared to Equation (5.93), it can be seen that the mechanism does not exactly deliver the correct relationship over the full range of steer angles. However, with careful design, only a very small error is present. Secondly, the fundamentals of Ackermann steering ignore the fact that all the tires will experience some slip angle as the vehicle corners, so the ideal relationship should depend on lateral acceleration as well as steer angle. As a result, most manufacturers will design the kinematics of their steering systems to deliver far less toe-out than called for by Ackermann, especially at small steer angles. The reasoning is that at higher speeds, where larger lateral acceleration occurs, the steer angles are generally much smaller. For low speed manœuvring, e.g., in a parking lot, there are more likely to be large steer angles and small slip angles, so more Ackermann effect is desirable.

5.5 Suspension Damping

It is quite common in the development of vehicle dynamic models to treat the suspension damping as linear, i.e., the force at the damper is assumed to be proportional to the stroke speed of the damper. In practice, this is almost never exactly true. There are a number of significant factors to consider here. First, most dampers in use today are based on a hydraulic principle. Oil is forced through an orifice in response to the motion of a piston to provide the resistance to motion. It is well known that the resulting hydraulic forces are not naturally linearly related to the flow speed through the orifice, but rather they tend to grow proportionally to the stroke speed raised to some power. As a result, most dampers use some sort of spring based relief valve, opening in response to the fluid pressure passing through it, to increase the size of the orifice and reduce the nonlinearity.

Secondly, the piston in the hydraulic cylinder must somehow be driven by a rod, which decreases the cross sectional area on one side of the piston relative to the other. This causes different flow rates of oil through the orifice when the piston is moving in opposing directions. Not only does this result in differing force characteristics in the two directions, but also, some sort of fluid reservoir must be built into the damper to accommodate the change in fluid volume that is required. Most dampers use two orifices, each fitted with one-way valves, to separate the design parameters that control the resistance to flow in each direction.

Finally, in practice, the actual damping rates that are desired in the two directions are not the same. When the vehicle is driven over a bump in the road, and the damper is moving in compression, it is preferable to minimize the resistance of the suspension to the motion of the wheel up and over the bump, rather than applying the large forces required to overcome the substantially larger inertia of the vehicle chassis. However, in the event of rebound, when the vehicle drives over a pothole, it is preferable to delay the extension of the suspension, to prevent the wheel from falling into the gap, only to be immediately be forced up again at the end. Instead, the vehicle relies on the large inertia of the chassis to prevent it from falling into the pothole, and lifts the wheel over the gap. As a consequence, most production vehicles have intentionally distinct compression and rebound damping behaviour.

One of the perhaps unexpected effects of the asymmetric damping curves is the tendency of the suspension to 'squat', or compress, when driving over an extended sequence of oscillatory disturbances. The displacement and velocity in the compression and rebound directions is equal and opposite, but the net force required is nonzero, due to the larger forces required in the rebound stroke. To provide this larger rebound force, the suspension settles into a state of vibration around a position that is slightly compressed relative to the static ride height, with the extra spring compression providing the additional rebound force. This has the unfortunate effect of reducing the ability of the suspension to absorb large disturbances, as a portion of the travel is already used by the preloading effect.

Problems

5.1 Consider a vehicle with the following parameters:
Mass: 2000 kg
Wheelbase ($a + b$): 2.3 m
Weight distribution (%f/r): 55/45
Anti-roll bar stiffness: front only, 5 kN/m (measured by deflecting one suspension while holding the other side fixed, this is in addition to the normal suspension stiffness)
Height of the centre of mass: 0.6 m (above the ground)
Track width: 1.4 m (left to right, same for both front and rear axles)

a) Assume the vehicle is in a steady state corner (i.e., the yaw rate is constant), where the lateral acceleration is 0.3 g to the right. Find the total lateral force provided by each axle (i.e., front and rear). Hint: think about the development of the bicycle model.
b) In order to accurately predict oversteer/understeer effects of suspension stiffnesses, a tire model more sophisticated than the simple linear model is necessary. Consider the tire model $Y = f(\alpha)(\mu Z - \kappa Z^2)$, where Y are the lateral forces, Z are the normal forces, and f is a function related only to the slip angle. Use values of $\mu = 0.9$ and $\kappa = 6.0 \times 10^{-5}$/N. Use this tire model to estimate the lateral forces at each tire. Hint: begin by computing the normal load on each tire. Assume that both front tires are operating at the same slip angle, as are both back tires.

5.2 One of the common design compromises that must be made in vehicle design is in suspension stiffness. Typically, a softer suspension helps to improve both ride and tire grip, but allows excessive body motions under cornering, braking, or acceleration. Consider the following specifications of a small off-road military truck.
Sprung mass: 1260 kg
Pitch moment of inertia: 1620 kg m^2
Wheelbase dimension: $a = 0.970$ m
Wheelbase dimension: $b = 1.047$ m
Centre of mass height: 0.570 m
Assume that identical suspension components are used on all four wheels.

a) Find the maximum acceleration before the rear wheel spins. Assume that the acceleration is not engine limited, that the tire has a coefficient of friction of $\mu = 0.75$, and that the vehicle is operating in rear wheel drive (RWD) mode, on level ground.
b) Find the longitudinal weight transfer that occurs under the maximum acceleration.
c) Find the suspension stiffness that results in a pitch angle of 5° under maximum acceleration. The pitch angle can be approximated by the expression:

$$\theta = \frac{z_r - z_f}{a + b} \tag{5.94}$$

where z is the suspension deflection, with compression taken as positive, rebound as negative, and equilibrium taken as zero pitch. Note the importance of units.
d) Find the resulting natural frequencies of the bounce/pitch model that would result from this choice of stiffness. Are they realistic?
e) Is there any way in which you could adjust the pitch frequency without changing the pitch angle under braking or acceleration? How and why would you do this?

5.3 Given a passenger car with the following parameters:
Mass: 2000 kg
Wheelbase: 2.3 m
Distance from the centre of mass to the rear axle: 1.2 m
Height of the centre of mass: 0.6 m (above the ground)
Rear suspension wheel rate (i.e., stiffness at the wheel): 18 kN/m (at each wheel)
Rear suspension motion ratio: 2:1 (i.e., the wheel moves twice the amount of the spring, treat as constant)
Rear wheel drive
Coefficient of friction: 0.8

Problem 5.3 Trailing arm suspension.

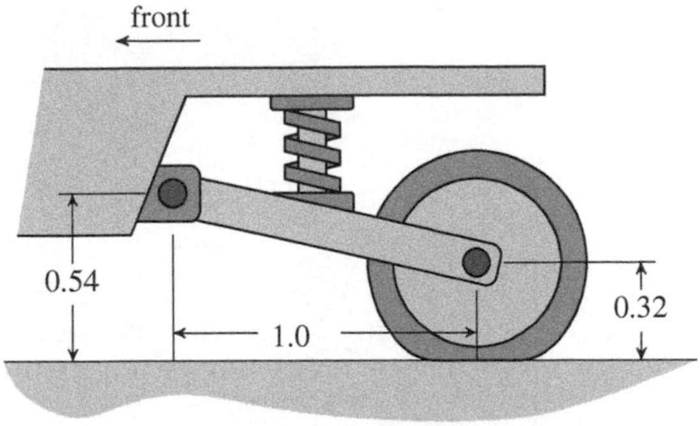

a) Determine the deflection of the rear suspension springs when the vehicle is at rest.
b) Determine the additional deflection of the rear suspension springs when traction limited maximum acceleration is achieved, ignoring kinematic effects of the rear suspension.
c) Assume the rear axle is a trailing arm as pictured in Problem 5.3, taking the dimensions as meters. Determine an estimate of the additional deflection of the rear suspension springs in the condition above, when including kinematic effects. Hint: the traction forces will cause jacking in the rear suspension, (sometimes referred to as *anti-squat*). Assume that moments are balanced around the axle pivot.

5.4 Suppose that you have taken a job with a recreational vehicle manufacturer. Your supervisor assigns you to a new project: a small passenger vehicle with tandem seating like a motorcycle, designed specifically for inner city travel. The proposal has a unique design feature; it is a 'reverse tricycle', i.e., a three-wheeled vehicle, with two front wheels, one driven rear wheel. Give a detailed explanation of any concerns you would have about the handling properties of such a vehicle. Points to consider: What does the bicycle model tell us about such a vehicle? What is fundamentally different about a tricycle, as opposed to a traditional four wheel design? What recommendations would you make about the location of the mass centre? What about the roll stiffness? Why?

References

1 Dixon, J.C., 2009. *Suspension Geometry and Computation.* John Wiley & Sons.

6

Multibody Dynamics

As the complexity of a vehicle model grows, the challenge of generating the equations of motion grows with it. This challenge reduces the practical limit for generating the equations of motion by hand to systems of relatively few bodies and degrees of freedom. For larger or more complex models, an automated procedure is required. The vehicle can be considered as a generic mechanical system, like the one shown in Figure 6.1, consisting of rigid bodies connected by various types of mechanical joints. The automatic generation and solution of the equations of motion of such a system is referred to as *multibody dynamics* (MBD).

6.1 Generating the Governing Equations

There are many techniques that can be used to find the equations of motion of a multibody system, e.g., Newton–Euler (the familiar sum of forces/moments), Lagrange's equation (an energy approach), D'Alembert's method (virtual work), Jourdain's principle (virtual power), Hamilton's equations, etc. This text is not meant to serve as an exhaustive study of the topic, and the interested reader is directed to one of the many excellent references available on the topic, e.g., Amirouche[1], Nikravesh[2], Shabana[3], Schiehlen[4].

In any case, no single method is clearly superior, as the best choice is often highly dependent on the problem itself. Regardless of the method chosen, it is important to remember that the resulting equations of motion must be equivalent. They will not necessarily be identical, as the choice of coordinates may differ between methods. A compromise that often appears is the choice between a model that has many simple equations, or a smaller but very complicated set.

Fundamentals of Vehicle Dynamics and Modelling: A Textbook for Engineers with Illustrations and Examples, First Edition. Bruce P. Minaker.

Companion website: www.wiley.com/go/minaker/vehicle-dynamics

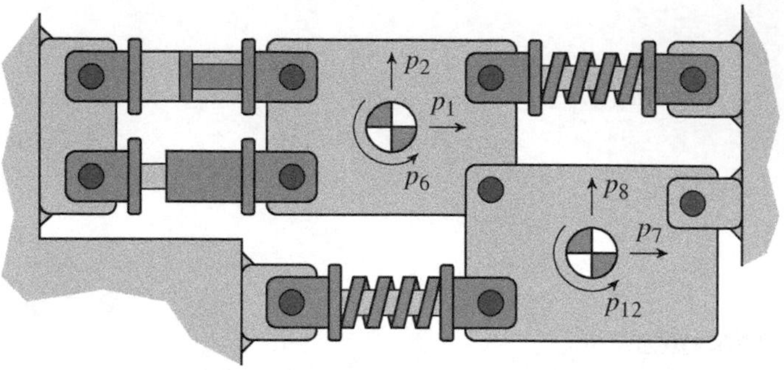

Figure 6.1 A generic multibody system.

In their most general form, the equations of motion of a constrained multibody system are a set of coupled nonlinear differential algebraic equations (DAEs). The complete set consists of three distinct parts:

- The *kinematic differential equations* relate the velocities (translational and rotational) to the rate of change of the positions and orientations. In many cases these are a straightforward time integration, but they can sometimes require a differential equation, particularly in the case of three dimensional rotation.
- The *Newton–Euler equations* describe the physical laws of motion that relate forces and moments to accelerations (again, translational and rotational).
- The *constraint equations*, typically an algebraic set that maintain the various relationships between the individual bodies due to the connectors, e.g., hinges or ball-joints.

Once the equations of motion have been formed, it is apparent that the next task is to find a solution to those equations. What is not so apparent is the method that should be applied to do so. The first question one might ask is whether it is possible to find an algebraic solution to the equations. Unfortunately, the cases in which it is possible to do so are quite few in number, and limited in application. The equations of motion are typically nonlinear, and sufficiently tangled that no closed form solution exists. Also, for most practical problems of interest, the equations are much more numerous than could reasonably be solved by hand, even if a closed form solution did exist. Fortunately, when using a modern MBD software tool, the details of these operations are typically invisible to the user.

This chapter details the development of a method of generating the linear equations of motion for large multibody models. The method is suitable for automation, and has been implemented in software. The equations of a number of sample vehicle systems will be generated, and the results will be discussed and interpreted.

6.1.1 Preliminary Definitions

In order to linearize the equations of motion, the concept of *variation* is introduced. Under a variation, a variable is assumed to change by a small amount relative to a given value. The δ operator is used to denote this small change, or variation.

$$\boldsymbol{x}^*(t) = \boldsymbol{x}(t) + \delta\boldsymbol{x}(t) \tag{6.1}$$

A thorough exploration of the calculus of variations is beyond the scope of this text, but it can be shown that the rules for calculating a variation are very similar to those for a differential, with the primary difference being that no change in the independent variable occurs during a variation. For example:

$$\boldsymbol{f}^*(\boldsymbol{x}, t) = \boldsymbol{f}(\boldsymbol{x}, t) + \delta\boldsymbol{f}(\boldsymbol{x}, t) = \boldsymbol{f}(\boldsymbol{x}, t) + \frac{\partial \boldsymbol{f}}{\partial \boldsymbol{x}}\delta\boldsymbol{x} \tag{6.2}$$

Note the absence of any δt term, even though there is time dependence in $\boldsymbol{f}$, as time is held fixed during the variation, (as opposed to a dt that would appear in a derivative).

Another important concept is the tilde notation (~) to represent a *skew symmetric* matrix. A skew symmetric matrix is defined for the vector $\boldsymbol{r} = x\hat{\mathbf{i}} + y\hat{\mathbf{j}} + z\hat{\mathbf{k}}$ as:

$$\tilde{\boldsymbol{r}} = \begin{bmatrix} 0 & -z & y \\ z & 0 & -x \\ -y & x & 0 \end{bmatrix} \tag{6.3}$$

A skew symmetric matrix is the negative of its own transpose ($\tilde{\boldsymbol{r}} = -\tilde{\boldsymbol{r}}'$), and multiplication by the skew symmetric matrix is equivalent to the vector *cross product*.

$$\tilde{\boldsymbol{r}}\boldsymbol{v} = \boldsymbol{r} \times \boldsymbol{v} = -\boldsymbol{v} \times \boldsymbol{r} = \tilde{\boldsymbol{v}}'\boldsymbol{r} \tag{6.4}$$

The skew symmetric matrix of a unit vector $\hat{\boldsymbol{u}}$ has the interesting property that:

$$\tilde{\boldsymbol{u}}\tilde{\boldsymbol{u}}' = \mathbf{I} - \hat{\boldsymbol{u}}\hat{\boldsymbol{u}}' \tag{6.5}$$

Note that the product of a vector with its own transpose is sometimes referred to as the *outer product*, and results in a matrix. This is in contrast to the *inner product*, more commonly known as the *dot product*, which produces a scalar result, and is written in the reverse order ($\hat{\boldsymbol{u}}'\hat{\boldsymbol{u}}$).

6.2 Definition of Coordinates

The process begins by assigning a set of coordinates to describe the motion of the bodies. A reference point, located at the centre of mass of each body is assigned six coordinates, three for position ($\boldsymbol{x}$) and three for orientation ($\boldsymbol{\theta}$).

The position vector defines the location of the mass centre relative to a ground fixed origin. Unfortunately, orientation is not so straightforward.

In this analysis, orientation is specified by a rotation through the angle θ around the unit vector $\hat{\boldsymbol{u}}$, where θ is the magnitude of the orientation vector, i.e., $\theta = |\boldsymbol{\theta}|$, and $\hat{\boldsymbol{u}}$ is the unit orientation vector, i.e., $\hat{\boldsymbol{u}} = \boldsymbol{\theta}/\theta$. However, one should note that in general, the individual components of this orientation vector *cannot* be taken as the rotation angle around each of the individual axes. In fact, one of the more challenging aspects of three dimensional motion is that rotation is not a vector quantity like translation. When a body experiences a sequence of translations, the order is unimportant in determining the final location. This is not the case for rotation, where the final orientation depends both on the order in which a sequence of individual rotations are applied, and on whether the consecutive rotations are applied around a ground fixed set of axes, or the body fixed axes, which change after each step in the sequence.

An alternate but commonly used convention to define orientation is known as *Euler angles*, where three rotations are assumed to occur in a specific predefined sequence. There are multiple choices available, but in vehicle dynamics, it is common to use the *yaw-pitch-roll* convention, where the yaw angle ψ around the ground fixed z axis occurs first, followed by the pitch angle θ around the resulting intermediate y axis, and finally the roll angle ϕ around the resulting body fixed x axis.

In general, vectors in misaligned reference frames can be related through the use of a 3×3 *transformation matrix* $\mathbf{R}$, sometimes simply referred to as a *rotation matrix*. In general, the rotation matrix is a function of the relative misalignment of the two reference frames. The form of the rotation matrix depends on the means by which the orientation is defined. Another of the challenges concerning three dimensional rotation appears in this definition. In some cases, differing sets of rotation angles can result in the same rotation matrix. This condition is referred to as a *singularity*. Changing to an alternate means of specifying orientation may result in the singularity occurring at a different orientation, but no set of three orientation coordinates exists that entirely avoids singularities. There are choices that use four orientation parameters that will avoid singularities, but these have the additional complication that they are no longer independent, and together they must satisfy a constraint in order to be a valid set.

In this analysis, the restriction that the angular motion is limited to small motions will be imposed. This is a requirement for the linearization process, but conveniently it also eliminates many of the issues associated with the selection of coordinates, and allows the orientation to be represented as a vector quantity.

$$\boldsymbol{\theta} = \begin{bmatrix} \phi & \theta & \psi \end{bmatrix}' \tag{6.6}$$

The yaw, pitch, and roll notation is maintained, and ϕ, θ, and ψ are the independent rotations around the x, y, and z axes respectively.

The position and orientation coordinates are written in a $6n \times 1$ combined vector as $\boldsymbol{p}$:

$$\boldsymbol{p} = \begin{bmatrix} \boldsymbol{x}'_1 & \boldsymbol{\theta}'_1 & \boldsymbol{x}'_2 & \boldsymbol{\theta}'_2 & \cdots & \boldsymbol{x}'_n & \boldsymbol{\theta}'_n \end{bmatrix}' \tag{6.7}$$

where n is the number of bodies in the system, excluding the ground. The positions and orientations of each body are defined in a fixed global reference frame, and are all taken as zero when the system is in its reference configuration. In contrast, the velocities ($\boldsymbol{v}$) and angular velocities ($\boldsymbol{\omega}$) of each body are defined in a body fixed rotating reference frame attached to the individual bodies. However, the orientation of the body fixed reference frames is such that they are initially aligned with the ground fixed frame when the system is in its reference configuration. The velocities and angular velocities are also written in a combined vector as $\boldsymbol{w}$.

$$\boldsymbol{w} = \begin{bmatrix} \boldsymbol{v}'_1 & \boldsymbol{\omega}'_1 & \boldsymbol{v}'_2 & \boldsymbol{\omega}'_2 & \cdots & \boldsymbol{v}'_n & \boldsymbol{\omega}'_n \end{bmatrix}' \tag{6.8}$$

6.3 Kinematic Differential Equations

The kinematic differential equations relate the velocity coordinates to the rate of change of the position coordinates. In the case of translation, if the position and velocity are defined in the same reference frame, then velocity is simply the time derivative of position. This relationship becomes more complicated when different reference frames are used for each of the two quantities, e.g., when positions are measured from a ground fixed reference, and velocities are measured from a body fixed reference.

One should also note that the treatment of orientation is a different matter than that of location. Angular position can be related to angular velocity by a direct time integration only in the case that the rotation is described by a single coordinate, i.e., it is constrained to occur around a fixed axis. When three dimensional angular velocity and multiple orientation angles are considered, a significantly more elaborate nonlinear differential equation is required to relate the rate of change of the orientation angles to the angular velocity.

The development begins by considering translation. The relationship between the ground fixed rate of change of location and the body fixed velocity is given by a rotation matrix:

$$\dot{\boldsymbol{x}} = \mathbf{R}(\boldsymbol{\theta})\boldsymbol{v} \tag{6.9}$$

where:

$$\mathbf{R} = \mathbf{I} + \sin\theta\tilde{\boldsymbol{u}} + (1 - \cos\theta)\tilde{\boldsymbol{u}}\tilde{\boldsymbol{u}} \tag{6.10}$$

Note the appearance of the skew symmetric matrix in the definition. The rotation matrix has the important property of *orthogonality*. The columns (and the rows) form a set of orthogonal unit vectors. The result is that the inverse of the rotation matrix is simply its transpose:

$$\mathbf{R}\mathbf{R}^{-1} = \mathbf{R}\mathbf{R}' = \mathbf{I} \tag{6.11}$$

Limiting the orientation to small rotations allows the rotation matrix to be simplified.

$$\mathbf{R} \approx \mathbf{I} + \tilde{\theta} = \begin{bmatrix} 1 & -\psi & \theta \\ \psi & 1 & -\phi \\ -\theta & \phi & 1 \end{bmatrix} \tag{6.12}$$

Once the rotation matrix has been simplified, its orthogonality only holds to a first order.

$$\mathbf{R}\mathbf{R}' = (\mathbf{I} + \tilde{\theta})(\mathbf{I} - \tilde{\theta}) = \mathbf{I} - \tilde{\theta}\tilde{\theta} \tag{6.13}$$

However, if the rotations are small, the $\tilde{\theta}\tilde{\theta}$ term will be negligible. To relate angular motions, a similar relationship is used:

$$\dot{\theta} = \mathbf{S}(\theta)\omega \tag{6.14}$$

where:

$$\mathbf{S} = \mathbf{I} + \frac{\theta}{2}\tilde{u} + \left(1 - \frac{\theta \sin\theta}{2(1 - \cos\theta)}\right)\tilde{u}\tilde{u} \tag{6.15}$$

For small rotation angles, the expression simplifies.

$$\mathbf{S} \approx \mathbf{I} + \frac{1}{2}\tilde{\theta} \tag{6.16}$$

A 6×6 matrix **P** used in the kinematic differential equations is defined in terms of **R** and **S**.

$$\mathbf{P} = \begin{bmatrix} \mathbf{R} & 0 \\ 0 & \mathbf{S} \end{bmatrix} \tag{6.17}$$

The kinematic differential equations become:

$$\begin{Bmatrix} \dot{x} \\ \dot{\theta} \end{Bmatrix} = \begin{bmatrix} \mathbf{R} & 0 \\ 0 & \mathbf{S} \end{bmatrix} \begin{Bmatrix} v \\ \omega \end{Bmatrix} \tag{6.18}$$

or:

$$\dot{p} = \mathbf{P}w \tag{6.19}$$

When the orientation angles are all zero, i.e., the frames are aligned, then **P** reduces to the identity matrix, and the positions and velocities are then momentarily related by a simple differentiation. The position relations from Equation (6.19) can be written as:

$$\dot{x} = \mathbf{R}v = v + \theta \times v \tag{6.20}$$

or, equivalently:

$$\dot{\boldsymbol{x}} + \boldsymbol{v} \times \boldsymbol{\theta} - \boldsymbol{v} = \mathbf{0} \tag{6.21}$$

In order to linearize the kinematic differential equations, the orientations are assumed to be small. If all the quantities are assumed to vary by some small amount around a fixed value, one can write:

$$\delta\dot{\boldsymbol{x}} + \tilde{\boldsymbol{v}}\delta\boldsymbol{\theta} - \tilde{\boldsymbol{\theta}}\delta\boldsymbol{v} - \delta\boldsymbol{v} = \mathbf{0} \tag{6.22}$$

If the system is assumed to stay near the configuration $\boldsymbol{\theta} = \mathbf{0}$, the position kinematic equations can be simplified.

$$\delta\dot{\boldsymbol{x}} + \tilde{\boldsymbol{v}}\delta\boldsymbol{\theta} - \delta\boldsymbol{v} = \mathbf{0} \tag{6.23}$$

Next, the process is repeated with the rotation equation.

$$\dot{\boldsymbol{\theta}} = \mathbf{S}\boldsymbol{\omega} = \boldsymbol{\omega} + \frac{1}{2}\boldsymbol{\theta} \times \boldsymbol{\omega} \tag{6.24}$$

$$\dot{\boldsymbol{\theta}} - \boldsymbol{\omega} + \boldsymbol{\omega} \times \frac{1}{2}\boldsymbol{\theta} = \mathbf{0} \tag{6.25}$$

$$\delta\dot{\boldsymbol{\theta}} - \delta\boldsymbol{\omega} + \frac{1}{2}\tilde{\boldsymbol{\omega}}\delta\boldsymbol{\theta} - \frac{1}{2}\tilde{\boldsymbol{\theta}}\delta\boldsymbol{\omega} = \mathbf{0} \tag{6.26}$$

The rotation equation is linearized around a point where both the orientation $\boldsymbol{\theta} = \mathbf{0}$ and the angular velocity of the reference frame $\boldsymbol{\omega} = \mathbf{0}$.

$$\delta\dot{\boldsymbol{\theta}} - \delta\boldsymbol{\omega} = \mathbf{0} \tag{6.27}$$

To accommodate those cases where a body has a constant nonzero angular velocity, the body can be allowed to rotate relative to the reference frame, but only if certain conditions apply. This point will be expanded in the following sections. Combining the linearized kinematic equations gives:

$$\begin{Bmatrix} \delta\dot{\boldsymbol{x}} \\ \delta\dot{\boldsymbol{\theta}} \end{Bmatrix} + \begin{bmatrix} \mathbf{0} & \tilde{\boldsymbol{v}} \\ \mathbf{0} & \mathbf{0} \end{bmatrix} \begin{Bmatrix} \delta\boldsymbol{x} \\ \delta\boldsymbol{\theta} \end{Bmatrix} - \begin{Bmatrix} \delta\boldsymbol{v} \\ \delta\boldsymbol{\omega} \end{Bmatrix} = \mathbf{0} \tag{6.28}$$

or, using a more compact notation:

$$\delta\dot{\boldsymbol{p}} + \mathbf{V}\delta\boldsymbol{p} - \delta\boldsymbol{w} = 0 \tag{6.29}$$

The result can now be presented as a first order linear differential equation in the variations.

$$\begin{bmatrix} \mathbf{I} & \mathbf{0} \end{bmatrix} \begin{Bmatrix} \delta\dot{\boldsymbol{p}} \\ \delta\dot{\boldsymbol{w}} \end{Bmatrix} + \begin{bmatrix} \mathbf{V} & -\mathbf{I} \end{bmatrix} \begin{Bmatrix} \delta\boldsymbol{p} \\ \delta\boldsymbol{w} \end{Bmatrix} = \mathbf{0} \tag{6.30}$$

The **V** matrix results from the linearization of the kinematic differential equations, and depends on the velocities at the point around which the kinematic differential equations are linearized. It contains the skew symmetric matrix of the constant linear velocities of the bodies, arranged in the upper right 3×3 sub-matrix of the set of 6×6 matrices arranged along the diagonal. All other entries are zero.

Example

Consider the case of a single body moving in the horizontal plane. If the body is moving with the standard velocities defined in its body fixed reference frame, the components in the ground frame can be found using the trigonometric relations shown.

$$\begin{Bmatrix} \dot{x} \\ \dot{y} \end{Bmatrix} = \begin{bmatrix} \cos\psi & -\sin\psi \\ \sin\psi & \cos\psi \end{bmatrix} \begin{Bmatrix} u \\ v \end{Bmatrix}$$

If the restriction of small angles is imposed, then the small angle relations ($\sin\theta \approx \theta$, $\cos\theta \approx 1$) can be substituted. The small angle relations are simply the Taylor series, with all but the first term discarded.

$$\begin{Bmatrix} \dot{x} \\ \dot{y} \end{Bmatrix} = \begin{bmatrix} 1 & -\psi \\ \psi & 1 \end{bmatrix} \begin{Bmatrix} u \\ v \end{Bmatrix} = \begin{Bmatrix} u - v\psi \\ v + u\psi \end{Bmatrix}$$

The variation is taken, and the fixed value of the heading angle is set to zero, i.e., $\psi = 0$. (Note that this does not imply $\delta\psi = 0$.)

$$\begin{Bmatrix} \delta\dot{x} \\ \delta\dot{y} \end{Bmatrix} = \begin{Bmatrix} \delta u - v\delta\psi \\ \delta v + u\delta\psi \end{Bmatrix}$$

Rearranging gives:

$$\begin{Bmatrix} \delta\dot{x} \\ \delta\dot{y} \end{Bmatrix} + \begin{bmatrix} v & -1 & 0 \\ -u & 0 & -1 \end{bmatrix} \begin{Bmatrix} \delta\psi \\ \delta u \\ \delta v \end{Bmatrix} = \begin{Bmatrix} 0 \\ 0 \end{Bmatrix}$$

or:

$$\begin{bmatrix} 1 & 0 & 0 & 0 & 0 & 0 \\ 0 & 1 & 0 & 0 & 0 & 0 \\ 0 & 0 & 1 & 0 & 0 & 0 \end{bmatrix} \begin{Bmatrix} \delta\dot{x} \\ \delta\dot{y} \\ \delta\dot{\psi} \\ \delta\dot{u} \\ \delta\dot{v} \\ \delta\dot{r} \end{Bmatrix} + \begin{bmatrix} 0 & 0 & v & -1 & 0 & 0 \\ 0 & 0 & -u & 0 & -1 & 0 \\ 0 & 0 & 0 & 0 & 0 & -1 \end{bmatrix} \begin{Bmatrix} \delta x \\ \delta y \\ \delta\psi \\ \delta u \\ \delta v \\ \delta r \end{Bmatrix} = \begin{Bmatrix} 0 \\ 0 \\ 0 \end{Bmatrix}$$

The result is a set of three first order linear differential equations in six variables. Clearly, an additional three equations would be required to complete the set such that a solution could be found.

6.4 Newton–Euler Equations

The next step in generating the equations of motion is to write the Newton–Euler equations of motion for each body; these equate the mass times the rate of change of velocity to the sum of forces. These forces are a function of the position and velocity of the bodies, and time. Here the translational and rotational equations of motion are grouped, so the descriptions are applied in a general sense, i.e., the mass includes both the mass and inertia terms, the velocity includes the linear and angular velocity, and the forces include both the forces and moments.

$$\mathbf{M}\dot{w} = \sum f(p, w, t) \tag{6.31}$$

The mass matrix $\mathbf{M}$ is formed by filling its diagonal with the mass and inertia values of each of the bodies in the system.

$$\mathbf{M} = \begin{bmatrix} m_1 & 0 & 0 & 0 & 0 & 0 & \\ 0 & m_1 & 0 & 0 & 0 & 0 & \\ 0 & 0 & m_1 & 0 & 0 & 0 & \\ 0 & 0 & 0 & I_{\mathrm{xx}_1} & -I_{\mathrm{xy}_1} & -I_{\mathrm{xz}_1} & \cdots \\ 0 & 0 & 0 & -I_{\mathrm{xy}_1} & I_{\mathrm{yy}_1} & -I_{\mathrm{yz}_1} & \\ 0 & 0 & 0 & -I_{\mathrm{xz}_1} & -I_{\mathrm{yz}_1} & I_{\mathrm{zz}_1} & \\ & & & \vdots & & & \ddots \end{bmatrix} \tag{6.32}$$

The forces are sorted into a number of different categories:

- The inertial forces f_{i} are the centripetal forces and gyroscopic moments that appear in the Newton–Euler equations; expressions for these terms are well known.
- The elastic forces f_{e} are generated by 'flexible' joints, e.g., a spring or bushing, and are modelled using a *constitutive equation*, where the forces or moments they produce will be related to their deflection or rate of deflection. One might argue that damping forces do not meet the precise definition of 'elastic', (i.e., deformation does not necessarily return to zero when the forces are removed) but in this context they are included nevertheless.
- The constraint forces f_{c} are those applied by 'rigid' connections, i.e., those that reduce the degrees of freedom of the system, e.g., a hinge or ball joint. The rigid connectors are modelled using algebraic constraint equations; expressions for the constraint forces will be developed in the discussion of the constraint equations.
- The applied forces f_{a} are anything external to the system, modelled as an arbitrary function of time, and are assumed to be chosen by the analyst.

The equation of motion can be rewritten as:

$$\mathbf{M}\dot{w} = \sum f_{\mathrm{i}} + \sum f_{\mathrm{e}} + \sum f_{\mathrm{c}} + \sum f_{\mathrm{a}} \tag{6.33}$$

To linearize the Newton–Euler equations, a variation is taken.

$$\mathbf{M}\delta\dot{w} = \sum \delta f_{\mathrm{i}} + \sum \delta f_{\mathrm{e}} + \sum \delta f_{\mathrm{c}} + \sum \delta f_{\mathrm{a}} \tag{6.34}$$

6.4.1 Inertial Forces

The nonlinear inertial terms are a result of the acceleration expressions; the components that involve velocity are gathered and brought to the other side of the expression. In order to deal with the equations of motion of systems including axisymmetric rotating bodies e.g., flywheels, where the orientation is not important, but where the angular velocity can influence the other motions, the concept of *spin* is introduced. The spin is an additional constant angular velocity, denoted $\boldsymbol{s}$, that is measured relative to the rotating reference frame, with the requirement that the spin motion does not affect the inertia matrix as measured in the rotating reference frame. The effect of the spin velocity appears in the expression for the gyroscopic moments.

$$\boldsymbol{f}_{\mathrm{i}} = \begin{Bmatrix} \boldsymbol{f}_{\mathrm{i}_1} \\ \vdots \\ \boldsymbol{f}_{\mathrm{i}_n} \end{Bmatrix} = -\begin{Bmatrix} \boldsymbol{\omega}_1 \times m_1 \boldsymbol{v}_1 \\ \boldsymbol{\omega}_1 \times \mathbf{I}_{\mathrm{G}_1}\{\boldsymbol{\omega}_1 + \boldsymbol{s}_1\} \\ \vdots \\ \boldsymbol{\omega}_n \times m_n \boldsymbol{v}_n \\ \boldsymbol{\omega}_n \times \mathbf{I}_{\mathrm{G}_n}\{\boldsymbol{\omega}_n + \boldsymbol{s}_n\} \end{Bmatrix} \tag{6.35}$$

The change in inertial forces can then be found as:

$$\delta\boldsymbol{f}_{\mathrm{i}} = -\begin{Bmatrix} \delta\boldsymbol{\omega}_1 \times m_1 \boldsymbol{v}_1 + \boldsymbol{\omega}_1 \times m_1 \delta\boldsymbol{v}_1 \\ \delta\boldsymbol{\omega}_1 \times \mathbf{I}_{\mathrm{G}_1}\{\boldsymbol{\omega}_1 + \boldsymbol{s}_1\} + \boldsymbol{\omega}_1 \times \mathbf{I}_{\mathrm{G}_1}\delta\boldsymbol{\omega}_1 \\ \vdots \end{Bmatrix} \tag{6.36}$$

Assuming that the angular velocity of the reference frame is zero at the point of linearization allows some simplification.

$$\delta\boldsymbol{f}_{\mathrm{i}} = -\begin{Bmatrix} \delta\boldsymbol{\omega}_1 \times m_1 \boldsymbol{v}_1 \\ \delta\boldsymbol{\omega}_1 \times \mathbf{I}_{\mathrm{G}_1}\boldsymbol{s}_1 \\ \vdots \end{Bmatrix} \tag{6.37}$$

Utilizing the tilde notation gives a compact result in a form that allows the inertial terms to be included with the traditional damping matrix. The inertial terms appear in the right 6 × 3 sub-matrix of each of the 6 × 6 matrices that appear along the diagonal.

$$\delta\boldsymbol{f}_{\mathrm{i}} = \begin{bmatrix} \mathbf{0} & m_1\tilde{\boldsymbol{v}}_1 & \\ \mathbf{0} & \widetilde{\mathbf{I}_{\mathrm{G}_1}\boldsymbol{s}_1} & \cdots \\ & \vdots & \ddots \end{bmatrix} \begin{Bmatrix} \delta\boldsymbol{v}_1 \\ \delta\boldsymbol{\omega}_1 \\ \vdots \end{Bmatrix} \tag{6.38}$$

The result can be presented in an even more compact notation by defining the inertial damping matrix $\mathbf{L}_i$, which includes the centripetal and gyroscopic effects.

$$\delta \boldsymbol{f}_i = -\mathbf{L}_i \delta \boldsymbol{w} \tag{6.39}$$

6.4.2 Elastic Forces

To illustrate the construction of the elastic stiffness and damping matrices, a deflection matrix **H** is introduced. The deflection matrix expresses the deflection of the elastic elements as a linear function of the motion coordinates. The deflection matrix is a function of the geometry only, and typically consists of some combination of unit vectors and 'radius' vectors; the former will find a component of the body's motion in the direction of the elastic element, and the later will give the motion of the elastic element due to rotation of the body. The term $\boldsymbol{\Delta}$ is used to represent the small deflections of the individual elastic elements. The deflection matrix is defined as:

$$\mathbf{H} = \frac{\partial \boldsymbol{\Delta}}{\partial \boldsymbol{p}} \tag{6.40}$$

For small motions, the deflections can be written as:

$$\boldsymbol{\Delta} = \mathbf{H}|_{p=0} \boldsymbol{p} \tag{6.41}$$

The elastic forces can be expressed as a function of the deflection of the individual elastic elements, like so:

$$\boldsymbol{f}_e = \mathbf{H}'([\mathbf{k}]\boldsymbol{\Delta} + \boldsymbol{\lambda}_e) = \mathbf{H}'([\mathbf{k}]\mathbf{H}\boldsymbol{p} + \boldsymbol{\lambda}_e) \tag{6.42}$$

where $[\mathbf{k}]$ is simply a diagonal matrix of all the individual elastic element stiffnesses:

$$[\mathbf{k}] = \begin{bmatrix} k_1 & 0 & 0 & \\ 0 & k_2 & 0 & \cdots \\ 0 & 0 & k_3 & \\ & \vdots & & \ddots \end{bmatrix} \tag{6.43}$$

and the term $\boldsymbol{\lambda}_e$ represents the preload in the elastic elements. It is assumed that the linearization is around the equilibrium position, where all the coordinates are zero. Note, however, that the elastic and constraint forces are not necessarily zero at this point; there may preload forces in the connectors. The elastic forces generate stiffness and damping matrices as follows:

$$\delta \boldsymbol{f}_e = \left.\frac{\partial \boldsymbol{f}_e}{\partial \boldsymbol{w}}\right|_{w=0} \delta \boldsymbol{w} + \left.\frac{\partial \boldsymbol{f}_e}{\partial \boldsymbol{p}}\right|_{p=0} \delta \boldsymbol{p} = -\mathbf{L}_e \delta \boldsymbol{w} - \mathbf{K}_e \delta \boldsymbol{p} \tag{6.44}$$

Differentiating gives the resulting expression for the stiffness matrix $\mathbf{K}_e$.

$$\frac{\partial \boldsymbol{f}_e}{\partial \boldsymbol{p}} = -\mathbf{K}_e = \mathbf{H}'[\mathbf{k}]\mathbf{H} + \frac{\partial \mathbf{H}'}{\partial \boldsymbol{p}}\boldsymbol{\lambda}_e \tag{6.45}$$

The first term is in the stiffness matrix is due only to the deflection of the elastic elements. The second term is known as a *tangent stiffness* matrix. The tangent stiffness matrix is due to changes in direction of the force in the elastic element, and not its magnitude. The vector $\boldsymbol{\lambda}_e$ is evaluated at equilibrium, so the tangent stiffness matrix requires that the preloads be determined prior to generation of the equations of motion, either using static equilibrium, or the force-deflection relations. If there is no preload in the mechanism at equilibrium, the tangent stiffness term evaluates to zero.

The expression for the elastic damping matrix $\mathbf{L}_e$ is not developed here, but it follows the same approach, i.e., using a matrix to relate the rate of deformation of the elastic elements to the velocities of the bodies. The resulting damping matrix has the same form as the first term in the stiffness matrix, as it is assumed there would be no preload in the dampers at equilibrium. The elastic damping matrix is summed with the inertial damping matrix to find the total damping matrix.

$$\mathbf{L} = \mathbf{L}_i + \mathbf{L}_e \tag{6.46}$$

Once the inertial and elastic forces have been linearized, the Newton–Euler equations can be written as:

$$\begin{bmatrix} \mathbf{0} & \mathbf{M} \end{bmatrix} \begin{Bmatrix} \delta\dot{\boldsymbol{p}} \\ \delta\dot{\boldsymbol{w}} \end{Bmatrix} + \begin{bmatrix} \mathbf{K}_e & \mathbf{L} \end{bmatrix} \begin{Bmatrix} \delta\boldsymbol{p} \\ \delta\boldsymbol{w} \end{Bmatrix} = \delta\boldsymbol{f}_c + \delta\boldsymbol{f}_a \tag{6.47}$$

Note that the sigma notation has been dropped from the equation, but summations are still assumed on the forces.

6.4.3 Linear Spring

The stiffness matrix for a single body connected to the ground by a single spring, as illustrated in Figure 6.2, is now developed. The result can be expanded for more complex cases in a similar fashion.

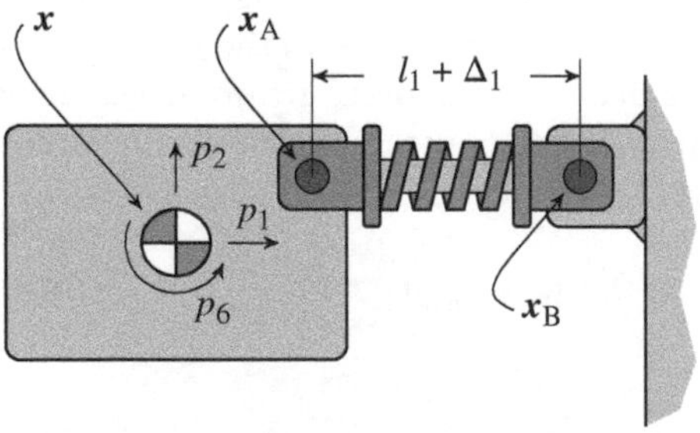

Figure 6.2 The deflection of the spring is a function of the motion of the mass centre. The resulting change in length and direction causes the change in the forces and moments acting on the body.

Now the stiffness matrix can be written by expanding expressions for the rate of change of force and moment with respect to translation and rotation of the body.

$$\mathbf{K}_e = -\frac{\partial \boldsymbol{f}_e}{\partial \boldsymbol{p}} = -\begin{bmatrix} \dfrac{\partial \boldsymbol{f}}{\partial \boldsymbol{x}} & \dfrac{\partial \boldsymbol{f}}{\partial \boldsymbol{\theta}} \\ \dfrac{\partial \boldsymbol{m}}{\partial \boldsymbol{x}} & \dfrac{\partial \boldsymbol{m}}{\partial \boldsymbol{\theta}} \end{bmatrix} \tag{6.48}$$

The location of the moving end of the spring is defined as $\boldsymbol{x}_A$, and the fixed end as $\boldsymbol{x}_B$. The location $\boldsymbol{x}_A$ can be found from the location $\boldsymbol{x}$ of the body, and the distance $\boldsymbol{r}$ of the end point A from the centre of mass. Note that $\boldsymbol{r}$ is defined as fixed in the rotating reference frame, and the rotation is assumed to be small.

$$\boldsymbol{x}_A = \boldsymbol{x} + \boldsymbol{r} + \boldsymbol{\theta} \times \boldsymbol{r} = \boldsymbol{x} + \boldsymbol{r} - \tilde{\boldsymbol{r}}\boldsymbol{\theta} \tag{6.49}$$

The rate of change of location of the end of the spring with respect to both the translation and rotation of the body can be found by differentiation. As expected, the motion of the end of the spring is the same as that of the mass centre when there is no rotation, and that the contribution of rotation depends on only $\boldsymbol{r}$.

$$\frac{\partial \boldsymbol{x}_A}{\partial \boldsymbol{x}} = \mathbf{I} \tag{6.50}$$

$$\frac{\partial \boldsymbol{x}_A}{\partial \boldsymbol{\theta}} = -\tilde{\boldsymbol{r}} = \tilde{\boldsymbol{r}}' \tag{6.51}$$

In the case of a two-point spring, the force vector can be defined in terms of its magnitude and direction. A simple linear spring force relationship is assumed, with the tension force depending on the length l, less the free unstretched length l_0, to define the magnitude.

$$f = k(l - l_0) \tag{6.52}$$

The length of the spring is determined from the locations of its end-points.

$$l = |\boldsymbol{x}_B - \boldsymbol{x}_A| \tag{6.53}$$

A unit vector acting along the spring defines the direction. The unit vector is expressed in global coordinates, as are the positions of the end-points of the spring.

$$\hat{\boldsymbol{u}} = \frac{\boldsymbol{x}_B - \boldsymbol{x}_A}{l} \tag{6.54}$$

Note that the force must be expressed in the body fixed coordinate system, as this is the system in which the equations of motion are written. The inverse rotation matrix is used to convert the force vector from the global frame to the body fixed frame. The magnitude of the force, being a scalar, can be factored out in front, and the rotation matrix can be expanded.

$$\boldsymbol{f} = \mathbf{R}'f\hat{\boldsymbol{u}} = f[\mathbf{I} - \tilde{\boldsymbol{\theta}}]\hat{\boldsymbol{u}} = f\hat{\boldsymbol{u}} + f\tilde{\boldsymbol{u}}\boldsymbol{\theta} \tag{6.55}$$

The first term in the stiffness matrix can now be expanded using the product rule.

$$\frac{\partial \boldsymbol{f}}{\partial \boldsymbol{x}} = \hat{\boldsymbol{u}}\frac{\partial f}{\partial \boldsymbol{x}} + f\frac{\partial \hat{\boldsymbol{u}}}{\partial \boldsymbol{x}} + \frac{\partial (f\tilde{\boldsymbol{u}})}{\partial \boldsymbol{x}}\boldsymbol{\theta} + f\tilde{\boldsymbol{u}}\frac{\partial \boldsymbol{\theta}}{\partial \boldsymbol{x}} \tag{6.56}$$

When considering the resulting expression, the third and fourth terms can be neglected. The third term will go to zero when the expression is evaluated at $\boldsymbol{\theta} = \mathbf{0}$, and the fourth will go to zero, as $\boldsymbol{\theta}$ is independent of $\boldsymbol{x}$. Considering the first term:

$$\frac{\partial f}{\partial \boldsymbol{x}} = k\frac{\partial l}{\partial \boldsymbol{x}_{\mathrm{A}}}\frac{\partial \boldsymbol{x}_{\mathrm{A}}}{\partial \boldsymbol{x}} = k\frac{\partial l}{\partial \boldsymbol{x}_{\mathrm{A}}} \tag{6.57}$$

Expanding the length expression as the root of the sum of squares of the vector components allows one to show that the rate of change of the length relative to the translation of one end is simply the negative unit vector along its length, expressed as a row vector.

$$\frac{\partial l}{\partial \boldsymbol{x}_{\mathrm{A}}} = -\hat{\boldsymbol{u}}' \tag{6.58}$$

Substitution gives:

$$\frac{\partial f}{\partial \boldsymbol{x}} = -k\hat{\boldsymbol{u}}' \tag{6.59}$$

Expanding the remaining derivative terms allows an algebraic expression for the first term in the stiffness matrix to be found.

$$\begin{aligned}\frac{\partial \hat{\boldsymbol{u}}}{\partial \boldsymbol{x}} &= \frac{\partial \hat{\boldsymbol{u}}}{\partial \boldsymbol{x}_{\mathrm{A}}}\frac{\partial \boldsymbol{x}_{\mathrm{A}}}{\partial \boldsymbol{x}} = \frac{\partial \hat{\boldsymbol{u}}}{\partial \boldsymbol{x}_{\mathrm{A}}}\\ &= -\frac{1}{l}\mathbf{I} + (\boldsymbol{x}_{\mathrm{B}} - \boldsymbol{x}_{\mathrm{A}})\frac{\hat{\boldsymbol{u}}'}{l^2} = -\frac{1}{l}[\mathbf{I} - \hat{\boldsymbol{u}}\hat{\boldsymbol{u}}'] = -\frac{1}{l}\tilde{\boldsymbol{u}}\tilde{\boldsymbol{u}}'\end{aligned} \tag{6.60}$$

Substitution gives:

$$\frac{\partial \boldsymbol{f}}{\partial \boldsymbol{x}} = -[\hat{\boldsymbol{u}}k\hat{\boldsymbol{u}}' + \tilde{\boldsymbol{u}}\frac{f}{l}\tilde{\boldsymbol{u}}'] \tag{6.61}$$

The second term in the stiffness matrix can now be expanded using the product rule.

$$\boldsymbol{f} = f\hat{\boldsymbol{u}} + f\tilde{\boldsymbol{u}}\boldsymbol{\theta} \tag{6.62}$$

$$\frac{\partial \boldsymbol{f}}{\partial \boldsymbol{\theta}} = \hat{\boldsymbol{u}}\frac{\partial f}{\partial \boldsymbol{\theta}} + f\frac{\partial \hat{\boldsymbol{u}}}{\partial \boldsymbol{\theta}} + \frac{\partial (f\tilde{\boldsymbol{u}})}{\partial \boldsymbol{\theta}}\boldsymbol{\theta} + f\tilde{\boldsymbol{u}} \tag{6.63}$$

Again, the third term in the expression can be neglected as it will go to zero when evaluated at $\boldsymbol{\theta} = \mathbf{0}$. The rate of change of the force relative to rotation can be found in terms of the force relative to translation.

$$\frac{\partial f}{\partial \boldsymbol{\theta}} = \frac{\partial f}{\partial \boldsymbol{x}_{\mathrm{A}}}\frac{\partial \boldsymbol{x}_{\mathrm{A}}}{\partial \boldsymbol{\theta}} = \frac{\partial f}{\partial \boldsymbol{x}}\tilde{\boldsymbol{r}}' \tag{6.64}$$

$$\frac{\partial \boldsymbol{u}}{\partial \boldsymbol{\theta}} = \frac{\partial \boldsymbol{u}}{\partial \boldsymbol{x}_{\mathrm{A}}}\frac{\partial \boldsymbol{x}_{\mathrm{A}}}{\partial \boldsymbol{\theta}} = \frac{\partial \boldsymbol{u}}{\partial \boldsymbol{x}}\tilde{\boldsymbol{r}}' \tag{6.65}$$

Substituting, and recognizing that when substituting $\boldsymbol{\theta} = \mathbf{0}$, that $\mathbf{R} = \mathbf{I}$, i.e., $f\tilde{\boldsymbol{u}} = \tilde{\boldsymbol{f}}$ gives:

$$\frac{\partial \boldsymbol{f}}{\partial \boldsymbol{\theta}} = \frac{\partial \boldsymbol{f}}{\partial \boldsymbol{x}}\tilde{\boldsymbol{r}}' + \tilde{\boldsymbol{f}} \tag{6.66}$$

Finally, the change of moment terms can be found.

$$\boldsymbol{m} = \boldsymbol{r} \times \boldsymbol{f} = \tilde{\boldsymbol{r}}\boldsymbol{f} \tag{6.67}$$

Recognizing that $\boldsymbol{r}$ is constant and substituting the expressions for force gives:

$$\frac{\partial \boldsymbol{m}}{\partial \boldsymbol{x}} = \tilde{\boldsymbol{r}}\frac{\partial \boldsymbol{f}}{\partial \boldsymbol{x}} \tag{6.68}$$

$$\frac{\partial \boldsymbol{m}}{\partial \boldsymbol{\theta}} = \tilde{\boldsymbol{r}}\frac{\partial \boldsymbol{f}}{\partial \boldsymbol{\theta}} = \tilde{\boldsymbol{r}}\frac{\partial \boldsymbol{f}}{\partial \boldsymbol{x}}\tilde{\boldsymbol{r}}' + \tilde{\boldsymbol{r}}\tilde{\boldsymbol{f}} \tag{6.69}$$

The stiffness matrix is now complete.

$$\mathrm{K_e} = \begin{bmatrix} [\hat{\boldsymbol{u}}k\hat{\boldsymbol{u}}' + \tilde{\boldsymbol{u}}\frac{f}{l}\tilde{\boldsymbol{u}}'] & [\hat{\boldsymbol{u}}k\hat{\boldsymbol{u}}' + \tilde{\boldsymbol{u}}\frac{f}{l}\tilde{\boldsymbol{u}}']\tilde{\boldsymbol{r}}' \\ sym & \tilde{\boldsymbol{r}}[\hat{\boldsymbol{u}}k\hat{\boldsymbol{u}}' + \tilde{\boldsymbol{u}}\frac{f}{l}\tilde{\boldsymbol{u}}']\tilde{\boldsymbol{r}}' \end{bmatrix} - \begin{bmatrix} 0 & \tilde{\boldsymbol{f}} \\ 0 & \tilde{\boldsymbol{r}}\tilde{\boldsymbol{f}} \end{bmatrix} \tag{6.70}$$

The stiffness matrix can be expanded as the sum of three components, two of which can be factored as below.

$$\mathrm{K_e} = \begin{bmatrix} \hat{\boldsymbol{u}} \\ \tilde{\boldsymbol{r}}\hat{\boldsymbol{u}} \end{bmatrix} k \begin{bmatrix} \hat{\boldsymbol{u}}' & (\tilde{\boldsymbol{r}}\hat{\boldsymbol{u}})' \end{bmatrix} + \begin{bmatrix} \tilde{\boldsymbol{u}} \\ \tilde{\boldsymbol{r}}\tilde{\boldsymbol{u}} \end{bmatrix} \frac{f}{l} \begin{bmatrix} \tilde{\boldsymbol{u}}' & (\tilde{\boldsymbol{r}}\tilde{\boldsymbol{u}})' \end{bmatrix} - \begin{bmatrix} 0 & \tilde{\boldsymbol{f}} \\ 0 & \tilde{\boldsymbol{r}}\tilde{\boldsymbol{f}} \end{bmatrix} \tag{6.71}$$

An examination of the three components reveals that only the first of the three terms depends on the stiffness of the spring, and that the second and third terms result from any nonzero force present in the spring when the body is in equilibrium, i.e., the spring preload terms appear in tangent stiffness matrices. The second term is due to any changes in direction of the spring as the body moves, and the third represents from the change of force and moment in the body fixed frame due to rotation of the body.

Example

Consider the case where a spring is aligned with the x axis, i.e., $\hat{\boldsymbol{u}} = [1, 0, 0]'$, with one end fixed to a rigid body a distance $\boldsymbol{r} = [0.1, 0.1, 0]'$ m from its mass centre, and the other end fixed to ground. The spring has a length $l = 0.25$ m, stiffness $k = 15000$ N/m, and preload $f = 2000$ N. Compute the associated stiffness matrix. The first term becomes:

$$\mathbf{K}_1 = \begin{bmatrix} 15000 & 0 & 0 & 0 & 0 & -4500 \\ 0 & 0 & 0 & 0 & 0 & 0 \\ 0 & 0 & 0 & 0 & 0 & 0 \\ 0 & 0 & 0 & 0 & 0 & 0 \\ 0 & 0 & 0 & 0 & 0 & 0 \\ -4500 & 0 & 0 & 0 & 0 & 1350 \end{bmatrix}$$

Note that the only terms are a stiffness due to translation in the x axis, a torsional stiffness around the z axis due to the y component of the radius vector, and the symmetric coupling terms. The second term is significantly more complex, and shows stiffness due to translation in both the y and z axes, along with three torsional stiffnesses, and four coupling terms.

$$\mathbf{K}_2 = \begin{bmatrix} 0 & 0 & 0 & 0 & 0 & 0 \\ 0 & 8000 & 0 & 0 & 0 & 2400 \\ 0 & 0 & 8000 & 2400 & -2400 & 0 \\ 0 & 0 & 2400 & 720 & -720 & 0 \\ 0 & 0 & -2400 & -720 & 720 & 0 \\ 0 & 2400 & 0 & 0 & 0 & 720 \end{bmatrix}$$

The third term shows only the effect of the changing orientation of the force relative to the body fixed reference frame. One should note that in order for the body to remain in equilibrium, another force or connector should also be acting, which may also generate terms in the same fashion, which can lead to some cancellation.

$$\mathbf{K}_3 = \begin{bmatrix} 0 & 0 & 0 & 0 & 0 & 0 \\ 0 & 0 & 0 & 0 & 0 & 2000 \\ 0 & 0 & 0 & 0 & -2000 & 0 \\ 0 & 0 & 0 & 0 & -600 & 0 \\ 0 & 0 & 0 & 0 & 600 & 0 \\ 0 & 0 & 0 & 0 & 0 & 600 \end{bmatrix}$$

The total stiffness matrix contribution for the single spring is the sum of the three terms.

$$\mathbf{K}_{\mathrm{e}} = \begin{bmatrix} 15000 & 0 & 0 & 0 & 0 & -4500 \\ 0 & 8000 & 0 & 0 & 0 & 4400 \\ 0 & 0 & 8000 & 2400 & -4400 & 0 \\ 0 & 0 & 2400 & 720 & -1320 & 0 \\ 0 & 0 & -2400 & -720 & 1320 & 0 \\ -4500 & 2400 & 0 & 0 & 0 & 2670 \end{bmatrix}$$

If the example is expanded to include a spring connecting two bodies, the resulting form of the expression is very similar, and given below:

$$\mathbf{K}_{\mathrm{e}} = \begin{bmatrix} \hat{\boldsymbol{u}} \\ \tilde{r}_1\hat{\boldsymbol{u}} \\ -\hat{\boldsymbol{u}} \\ -\tilde{r}_2\hat{\boldsymbol{u}} \end{bmatrix} k \left[\hat{\boldsymbol{u}}' \;\; (\tilde{r}_1\hat{\boldsymbol{u}})' \;\; -\hat{\boldsymbol{u}}' \;\; -(\tilde{r}_2\hat{\boldsymbol{u}})'\right]$$

$$+ \begin{bmatrix} \tilde{\boldsymbol{u}} \\ \tilde{r}_1\tilde{\boldsymbol{u}} \\ -\tilde{\boldsymbol{u}} \\ -\tilde{r}_2\tilde{\boldsymbol{u}} \end{bmatrix} \frac{f}{l} \left[\tilde{\boldsymbol{u}}' \;\; (\tilde{r}_1\tilde{\boldsymbol{u}})' \;\; -\tilde{\boldsymbol{u}}' \;\; -(\tilde{r}_2\tilde{\boldsymbol{u}})'\right] - \begin{bmatrix} 0 & \tilde{f}_{\mathrm{g}} & 0 & 0 \\ 0 & \tilde{r}_1\tilde{f}_{\mathrm{g}} & 0 & 0 \\ 0 & 0 & 0 & -\tilde{f}_{\mathrm{g}} \\ 0 & 0 & 0 & -\tilde{r}_2\tilde{f}_{\mathrm{g}} \end{bmatrix} \tag{6.72}$$

6.5 Constraint Equations

Once the Newton–Euler equations and kinematic differential equations are assembled, the final stage is the inclusion of the constraints. Most multibody systems of interest include some type of joint or connector that does not deform elastically, and eliminates some of the *degrees of freedom* (DOF) from the system. The degrees of freedom of a mechanism refers to the number of independent motions that can occur. An unconstrained body has six degrees of freedom, translation and rotation in each of the x, y, and z directions. The number of degrees of freedom of a mechanism is then $6n - m$, where n is the number of bodies, and m is the number of constraints. Each type of mechanical connector will add a number of constraints. Each degree of freedom that is removed has the appropriate motions enforced by a corresponding constraint force.

Consider a spherical or ball joint connection. At the point where it connects the two bodies, the translational motion of the two bodies must match in all three directions, but no restrictions are imposed on the rotational motion.

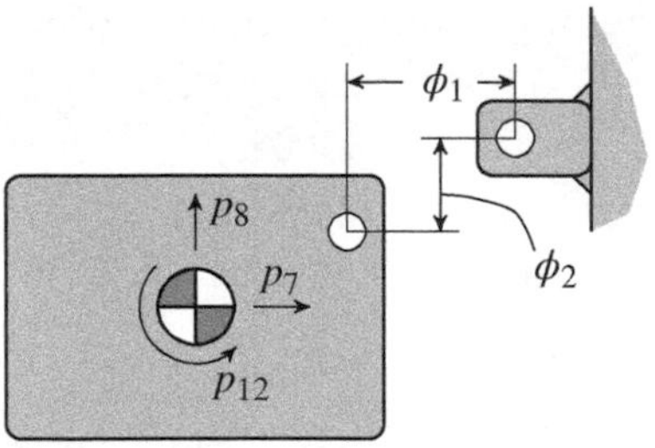

Figure 6.3 The constraint equations are typically written as a quantity that should go to zero, e.g., the vector distance between two connected points.

The spherical joint imposes three constraints, so two bodies moving in space, attached by a spherical joint would have nine degrees of freedom. Three constraint forces would act at the joint, with equal and opposite reactions between the two bodies.

The constraints are expressed as a set of algebraic equations that state some fixed relationship must be maintained by the positions, velocities, and time.

$$\boldsymbol{\phi}(\boldsymbol{p}, \boldsymbol{w}, t) = \boldsymbol{0} \tag{6.73}$$

Typically, the constraint equations are formulated such that they equal zero when the constraints are satisfied. In the case of the spherical joint, the vector $\boldsymbol{\phi}$ would represent the vector from the centre of the ball joint to the centre of the corresponding cup, computed from the location and orientation of each of the two adjoining bodies, as shown in Figure 6.3.

The constraints on a system can be categorized as *holonomic* or *nonholonomic*. The holonomic constraints can be expressed without any dependence on the velocities of the bodies. The holonomic constraints are further subcategorized as *rheonomic*, to indicate their explicit time dependence, or as *scleronomic*, if they are independent of time. In general, the equations of motion of nonholonomic systems are more challenging to produce; similarly rheonomic systems are generally more complicated than scleronomic systems. Fortunately, many common mechanical connections, such as the spherical joint, can be modelled using scleronomic constraints. However, it is not uncommon for vehicle systems to include nonholonomic constraints. A notable example is a wheel rolling without lateral slip. The general form of the scleronomic algebraic constraint equations is:

$$\boldsymbol{\phi}(\boldsymbol{p}) = \boldsymbol{0} \tag{6.74}$$

Each row of the $\boldsymbol{\phi}$ vector represents a single constraint, so most mechanical connectors will require more than one row. The constraint equations can be linearized by considering the motions to be small. Taking a variation gives:

$$\delta\boldsymbol{\phi} = \frac{\partial \boldsymbol{\phi}}{\partial \boldsymbol{p}} \delta\boldsymbol{p} = \boldsymbol{0} \tag{6.75}$$

The partial derivative of the constraint equations with respect to the coordinates is the constraint *Jacobian* matrix.

$$\mathrm{J_h} = \frac{\partial \boldsymbol{\phi}}{\partial \boldsymbol{p}} \tag{6.76}$$

The deflection of the rigid elements can then be expressed as a linear function of the motion coordinates, and must be zero.

$$\mathrm{J_h} \delta \boldsymbol{p} = \mathbf{0} \tag{6.77}$$

The nonholonomic constraint equations can be written is a similar fashion that allows their inclusion, as long as there is no explicit time dependency.

$$\mathrm{J_{nh}} \delta \boldsymbol{w} = \mathbf{0} \tag{6.78}$$

The $\mathrm{J_h}$ and $\mathrm{J_{nh}}$ matrices represent the holonomic and nonholonomic constraint Jacobians respectively. Once the constraint equations are formed, an expression for the corresponding constraint forces can be written as:

$$\boldsymbol{f}_{\mathrm{c}} = -\left[\mathrm{J}'_{\mathrm{h}} \;\; \mathrm{J}'_{\mathrm{nh}}\right] \boldsymbol{\lambda} = -\mathrm{J}' \boldsymbol{\lambda} \tag{6.79}$$

where $\boldsymbol{\lambda}$ is known as a *Lagrange multiplier*. This equation stems from the fact that a motion in a particular direction can only be resisted by a force in the same direction. The Lagrange multiplier scales a vector in each direction by some unknown amount. In the example of the spherical constraint, the three unknown Lagrange multipliers would be the three forces in the ball joint in the global reference frame. The Jacobian matrix would serve to transform them into the appropriate body fixed reference frame, and compute the resulting moment around the mass centre as well. The unknown values of the Lagrange multipliers can be found during the solution process. The constraint forces are linearized by taking a variation.

$$\begin{aligned} \delta \boldsymbol{f}_{\mathrm{c}} &= -\delta \mathrm{J}' \boldsymbol{\lambda} - \mathrm{J}' \delta \boldsymbol{\lambda} \\ &= -\frac{\partial J'}{\partial \boldsymbol{p}} \delta \boldsymbol{p} \boldsymbol{\lambda} - \mathrm{J}' \delta \boldsymbol{\lambda} \end{aligned} \tag{6.80}$$

The constraints generate another tangent stiffness matrix in the event that there are any preload forces carried when the system is in equilibrium. It is defined as:

$$\mathrm{K_c} = \frac{\partial \mathrm{J}'}{\partial \boldsymbol{p}} \boldsymbol{\lambda} \tag{6.81}$$

so the constraint forces can be expressed as:

$$\delta \boldsymbol{f}_{\mathrm{c}} = -\mathrm{K_c} \delta \boldsymbol{p} - \mathrm{J}' \delta \boldsymbol{\lambda} \tag{6.82}$$

The tangent stiffness term can be added to the elastic stiffness matrix to find the total stiffness matrix.

$$\mathrm{K} = \mathrm{K_e} + \mathrm{K_c} \tag{6.83}$$

The kinematic differential equations, Equation (6.29), and the Newton–Euler equations, Equation (6.47), can now be combined in a common first order set. The constraint forces are substituted in, with the remaining terms appearing on the right hand side.

$$\begin{bmatrix} \mathbf{I} & \mathbf{0} \\ \mathbf{0} & \mathbf{M} \end{bmatrix} \begin{Bmatrix} \delta \dot{\boldsymbol{p}} \\ \delta \dot{\boldsymbol{w}} \end{Bmatrix} + \begin{bmatrix} \mathbf{V} & -\mathbf{I} \\ \mathbf{K} & \mathbf{L} \end{bmatrix} \begin{Bmatrix} \delta \boldsymbol{p} \\ \delta \boldsymbol{w} \end{Bmatrix} = \begin{Bmatrix} \mathbf{0} \\ -\mathbf{J}' \delta \boldsymbol{\lambda} + \delta \boldsymbol{f}_{\mathrm{a}} \end{Bmatrix} \tag{6.84}$$

6.5.1 Spherical Joint

The linear velocity constraints and the tangent stiffness matrix for a spherical joint are developed. The joint is treated as a point that has a common velocity when expressed in the global frame. Note that while the linear velocity constraint would suffice to describe a spherical joint, other types, e.g., a hinge or revolute joint requires additional constraints on angular velocity. The velocity constraint is:

$$\mathbf{R}_1(\boldsymbol{v}_1 + \boldsymbol{\omega}_1 \times \boldsymbol{r}_1) - \mathbf{R}_2(\boldsymbol{v}_2 + \boldsymbol{\omega}_2 \times \boldsymbol{r}_2) = \mathbf{0} \tag{6.85}$$

$$\mathbf{R}_1 \boldsymbol{v}_1 - \mathbf{R}_1 \tilde{\boldsymbol{r}}_1 \boldsymbol{\omega}_1 - \mathbf{R}_2 \boldsymbol{v}_2 + \mathbf{R}_2 \tilde{\boldsymbol{r}}_2 \boldsymbol{\omega}_2 = \mathbf{0} \tag{6.86}$$

The equations are rewritten in matrix-vector form.

$$\begin{bmatrix} \mathbf{R}_1 & -\mathbf{R}_1 \tilde{\boldsymbol{r}}_1 & -\mathbf{R}_2 & \mathbf{R}_2 \tilde{\boldsymbol{r}}_2 \end{bmatrix} \begin{Bmatrix} \boldsymbol{v}_1 \\ \boldsymbol{\omega}_1 \\ \boldsymbol{v}_2 \\ \boldsymbol{\omega}_2 \end{Bmatrix} = \mathbf{0} \tag{6.87}$$

The equation is now in the form $\mathbf{J}\boldsymbol{w} = \mathbf{0}$, so this matrix $\mathbf{J}$ (the constraint Jacobian) can be used to write the constraint forces in terms of the Lagrange multipliers, as $\boldsymbol{f}_{\mathrm{c}} = \mathbf{J}'\boldsymbol{\lambda}$. In this case, the $\boldsymbol{\lambda}$ vector represents the force vector carried by the joint, expressed in the global reference frame, while $\boldsymbol{f}_1$, $\boldsymbol{m}_1$, etc., are the forces and moments around the respective centres of mass, in each local frame. Recall that the transpose of a matrix product can be found using the identity $(\mathbf{XY})' = \mathbf{Y}'\mathbf{X}'$.

$$\begin{bmatrix} \mathbf{R}'_1 \\ \tilde{\boldsymbol{r}}_1 \mathbf{R}'_1 \\ -\mathbf{R}'_2 \\ -\tilde{\boldsymbol{r}}_2 \mathbf{R}'_2 \end{bmatrix} \boldsymbol{\lambda} = \begin{Bmatrix} \boldsymbol{f}_1 \\ \boldsymbol{m}_1 \\ \boldsymbol{f}_2 \\ \boldsymbol{m}_2 \end{Bmatrix} \tag{6.88}$$

Next, the rotation matrices are expanded.

$$\begin{bmatrix} \mathbf{I}-\tilde{\theta}_1 \\ \tilde{r}_1-\tilde{r}_1\tilde{\theta}_1 \\ -\mathbf{I}+\tilde{\theta}_2 \\ -\tilde{r}_2+\tilde{r}_2\tilde{\theta}_2 \end{bmatrix} \lambda = \begin{Bmatrix} \boldsymbol{f}_1 \\ \boldsymbol{m}_1 \\ \boldsymbol{f}_2 \\ \boldsymbol{m}_2 \end{Bmatrix} \tag{6.89}$$

Recognizing that in this case the Lagrange multipliers are just the forces acting at the joint (in global coordinates) allows λ to be replaced with $\boldsymbol{f}_g$. Reversing the order of the cross product allows an expression for $\boldsymbol{f}_1$.

$$\boldsymbol{f}_1 = (\mathbf{I}-\tilde{\theta}_1)\boldsymbol{f}_g = \boldsymbol{f}_g + \tilde{\boldsymbol{f}}_g\boldsymbol{\theta}_1 \tag{6.90}$$

The stiffness term is now found by differentiation.

$$\frac{\partial \boldsymbol{f}_1}{\partial \boldsymbol{\theta}_1} = \tilde{\boldsymbol{f}}_g \tag{6.91}$$

Similarly, an expression is found for the moments.

$$\boldsymbol{m}_1 = (\tilde{r}_1 - \tilde{r}_1\tilde{\theta}_1)\boldsymbol{f}_g = \tilde{r}_1\boldsymbol{f}_g + \tilde{r}_1\tilde{\boldsymbol{f}}_g\boldsymbol{\theta}_1 \tag{6.92}$$

Differentiation gives the angular stiffness.

$$\frac{\partial \boldsymbol{m}_1}{\partial \boldsymbol{\theta}_1} = \tilde{r}_1\tilde{\boldsymbol{f}}_g \tag{6.93}$$

The results are summarized in matrix form.

$$\mathbf{K}_c = -\begin{bmatrix} 0 & \tilde{\boldsymbol{f}}_g & 0 & 0 \\ 0 & \tilde{r}_1\tilde{\boldsymbol{f}}_g & 0 & 0 \\ 0 & 0 & 0 & -\tilde{\boldsymbol{f}}_g \\ 0 & 0 & 0 & -\tilde{r}_2\tilde{\boldsymbol{f}}_g \end{bmatrix} \tag{6.94}$$

A examination of the resulting matrix yields some insight. There is a strong similarity to the terms in Equation (6.72), the elastic stiffness matrix for a spring. By examining the force terms, one can see that $\delta\boldsymbol{f} = -\tilde{\boldsymbol{f}}_g\delta\boldsymbol{\theta} = \delta\boldsymbol{\theta}\times\boldsymbol{f}_g$, i.e., the change in force is due simply to the relative change in direction, calculated using a cross product. The moment term is similar; the change in moment is the constant radius vector crossed against the change in force.

Note that in these differentiations, the Lagrange multipliers are treated as constants. This can be explained by the particular formulation of the equations of motion that is used in Equation (6.84). A term containing the variation of the Lagrange multipliers is present in the equations of motion. During the final stage of the process, when combining the equations of motion with the constraint equations, the Lagrange multipliers term will be eliminated from the equations, so only the first term needs to be considered.

The development for a hinge or revolute joint follows is similar to that of the spherical joint; the resulting tangent stiffness matrix is given below:

$$\mathrm{K_c} = -\begin{bmatrix} 0 & \tilde{\boldsymbol{f}}_{\mathrm{g}} & 0 & 0 \\ 0 & \tilde{\boldsymbol{r}}_1\tilde{\boldsymbol{f}}_{\mathrm{g}} - \tilde{\boldsymbol{u}}\tilde{\boldsymbol{u}}\tilde{\boldsymbol{m}}_{\mathrm{g}} & 0 & 0 \\ 0 & 0 & 0 & -\tilde{\boldsymbol{f}}_{\mathrm{g}} \\ 0 & 0 & 0 & -\tilde{\boldsymbol{r}}_2\tilde{\boldsymbol{f}}_{\mathrm{g}} + \tilde{\boldsymbol{u}}\tilde{\boldsymbol{u}}\tilde{\boldsymbol{m}}_{\mathrm{g}} \end{bmatrix} \tag{6.95}$$

Here $\boldsymbol{f}_{\mathrm{g}}$ and $\boldsymbol{m}_{\mathrm{g}}$ are the preload force and moment carried by the joint, $\boldsymbol{r}$ is the location relative to the centre of mass, and $\hat{\boldsymbol{u}}$ is a unit vector that defines the axis of rotation.

6.6 State Space Form

The remaining step is to reduce the equations to a set of *minimal coordinates*, i.e., the size of the state vector should be equal to the number of degrees of freedom. The mechanism used to do so will conveniently provide a means to simultaneously remove the unknown Lagrange multipliers. It is important to note that when the equations are written in a first order form, the displacements and velocities are independent. As a result, a constraint on the positions must also be applied to the velocities. Once the constraint equations have been linearized, the Jacobian matrix can be treated as a constant, and differentiation with respect to time gives:

$$\mathrm{J_h}\delta\dot{\boldsymbol{p}} = \mathbf{0} \tag{6.96}$$

and:

$$\mathrm{J_h}\delta\ddot{\boldsymbol{p}} = \mathbf{0} \tag{6.97}$$

The kinematic differential equation, Equation (6.29), can be substituted into the resulting expression to give:

$$\mathrm{J_h}\delta\boldsymbol{w} - \mathrm{J_h}\mathrm{V}\delta\boldsymbol{p} = \mathbf{0} \tag{6.98}$$

and:

$$\mathrm{J_h}\delta\dot{\boldsymbol{w}} - \mathrm{J_h}\mathrm{V}\delta\dot{\boldsymbol{p}} = \mathbf{0} \tag{6.99}$$

The resulting expressions are gathered and assembled in a complete constraint matrix as follows:

$$\begin{bmatrix} \mathrm{J_h} & 0 \\ -\mathrm{J_h}\mathrm{V} & \mathrm{J_h} \\ 0 & \mathrm{J_{nh}} \end{bmatrix} \begin{bmatrix} \delta\dot{\boldsymbol{p}} & \delta\boldsymbol{p} \\ \delta\dot{\boldsymbol{w}} & \delta\boldsymbol{w} \end{bmatrix} = \begin{bmatrix} 0 & 0 \\ 0 & 0 \\ 0 & 0 \end{bmatrix} \tag{6.100}$$

In order to reduce the equations to a minimal set of coordinates, an *orthogonal complement* of the Jacobian is used. The complement is a matrix **T** defined such that its columns are perpendicular to the rows of the Jacobian. More precisely, it is an orthonormal basis for the null space of the Jacobian. As a result of the definition, the product of the complement and the Jacobian is zero.

$$\begin{bmatrix} \mathbf{J}_{\mathrm{h}} & \mathbf{0} \\ -\mathbf{J}_{\mathrm{h}}\mathbf{V} & \mathbf{J}_{\mathrm{h}} \\ \mathbf{0} & \mathbf{J}_{\mathrm{nh}} \end{bmatrix} \mathbf{T} = \mathbf{0} \tag{6.101}$$

The complement matrix is not unique. For example, consider the case of a single vector in three dimensions. The null space is the plane that is normal to that vector. If two perpendicular vectors in the normal plane are sought, an infinite number of choices exist, distinguished by a rotation around the normal vector. There are several methods for calculating a suitable choice, e.g., the null space can be found using singular value decomposition. A new coordinate vector $\boldsymbol{x}$ is defined using the complement. It should be noted that this vector is now in nonphysical coordinates, i.e., the values may not directly correspond to the physical motions, although in some cases, a subset of the physical coordinates may be suitable. In any case, the physical coordinates can be recovered using the complement matrix.

$$\mathbf{T}\boldsymbol{x} = \begin{Bmatrix} \delta \boldsymbol{p} \\ \delta \boldsymbol{w} \end{Bmatrix} \tag{6.102}$$

As a result of this definition, any possible selection of the coordinate vector must satisfy the constraint equations.

$$\begin{bmatrix} \mathbf{J}_{\mathrm{h}} & \mathbf{0} \\ -\mathbf{J}_{\mathrm{h}}\mathbf{V} & \mathbf{J}_{\mathrm{h}} \\ \mathbf{0} & \mathbf{J}_{\mathrm{nh}} \end{bmatrix} \mathbf{T}\boldsymbol{x} = \begin{bmatrix} \mathbf{J}_{\mathrm{h}} & \mathbf{0} \\ -\mathbf{J}_{\mathrm{h}}\mathbf{V} & \mathbf{J}_{\mathrm{h}} \\ \mathbf{0} & \mathbf{J}_{\mathrm{nh}} \end{bmatrix} \begin{Bmatrix} \delta \boldsymbol{p} \\ \delta \boldsymbol{w} \end{Bmatrix} = \mathbf{0} \tag{6.103}$$

In a similar manner, find the complement matrix **U** such that:

$$\mathbf{U} \begin{bmatrix} \mathbf{J}'_{\mathrm{h}} & \mathbf{0} & \mathbf{0} \\ \mathbf{0} & \mathbf{J}'_{\mathrm{h}} & \mathbf{J}'_{\mathrm{nh}} \end{bmatrix} = \mathbf{0} \tag{6.104}$$

The new coordinates can be substituted into the equations of motion, Equation (6.84). The premultiplication by the orthogonal complement reduces the number of equations to match the number of coordinates, and eliminates the remaining constraint force terms from the equations. Once the equations have been reduced to minimal coordinates, the result will be a set of descriptor form, first order linear equations. These can be easily converted to standard form if the mass matrix is well conditioned; if it is not well conditioned (caused by e.g., the inclusion of 'massless' bodies), the conversion is more challenging, but still possible, using a singular value decomposition based technique.

$$\mathbf{U}\begin{bmatrix}\mathbf{I} & \mathbf{0}\\ \mathbf{0} & \mathbf{M}\end{bmatrix}\mathbf{T}\dot{\boldsymbol{x}}+\mathbf{U}\begin{bmatrix}\mathbf{V} & -\mathbf{I}\\ \mathbf{K} & \mathbf{L}\end{bmatrix}\mathbf{T}\boldsymbol{x}=\mathbf{U}\begin{Bmatrix}\mathbf{0}\\ \delta\boldsymbol{f}_{\mathrm{a}}\end{Bmatrix} \tag{6.105}$$

Then in order to simplify, the following definitions are made:

$$\mathbf{A}=-\mathbf{U}\begin{bmatrix}\mathbf{V} & -\mathbf{I}\\ \mathbf{K} & \mathbf{L}\end{bmatrix}\mathbf{T} \tag{6.106}$$

$$\mathbf{E}=\mathbf{U}\begin{bmatrix}\mathbf{0} & \mathbf{I}\\ \mathbf{0} & \mathbf{M}\end{bmatrix}\mathbf{T} \tag{6.107}$$

The terms defined above are substituted directly into the equations of motion. The applied forces are assumed to be chosen by the analyst as a function of time, and are included as inputs through the **B** matrix. In some cases, this matrix can be determined simply by selection of the appropriate columns of **U**, in more complicated systems, an approach similar to the Jacobian matrix can be used to compute it.

$$\mathbf{B}\boldsymbol{u}=\mathbf{U}\begin{Bmatrix}\mathbf{0}\\ \delta\boldsymbol{f}_{\mathrm{a}}\end{Bmatrix} \tag{6.108}$$

Substituting gives:

$$\mathbf{E}\dot{\boldsymbol{x}}=\mathbf{A}\boldsymbol{x}+\mathbf{B}\boldsymbol{u} \tag{6.109}$$

Typically the equations of motion are supplemented with an output equation of the form:

$$\boldsymbol{y}=\mathbf{C}\boldsymbol{x}+\mathbf{D}\boldsymbol{u} \tag{6.110}$$

Here the matrices **C** and **D** are chosen to provide the output information of interest, e.g., setting $\mathbf{C}=\mathbf{T}$ and $\mathbf{D}=\mathbf{0}$ implies that $\boldsymbol{y}=\mathbf{T}\boldsymbol{x}=[\delta\boldsymbol{p}'\delta\boldsymbol{w}']'$, i.e., the outputs are the full set of physical coordinates. Alternatively, choosing $\mathbf{C}=\mathbf{TA}$ and $\mathbf{D}=\mathbf{TB}$ implies that $\boldsymbol{y}=\mathbf{T}\dot{\boldsymbol{x}}=[\delta\dot{\boldsymbol{p}}'\delta\dot{\boldsymbol{w}}']'$, i.e., the outputs are the rate of change of the physical coordinates. Combing the two equations gives the well known descriptor form of the state space equations.

$$\begin{bmatrix}\mathbf{E} & \mathbf{0}\\ \mathbf{0} & \mathbf{I}\end{bmatrix}\begin{Bmatrix}\dot{\boldsymbol{x}}\\ \boldsymbol{y}\end{Bmatrix}=\begin{bmatrix}\mathbf{A} & \mathbf{B}\\ \mathbf{C} & \mathbf{D}\end{bmatrix}\begin{Bmatrix}\boldsymbol{x}\\ \boldsymbol{u}\end{Bmatrix} \tag{6.111}$$

The techniques outlined above have been compiled into a software suite capable of automatically generating the linearized equations of motion of a rigid multibody system. Collectively referred to as 'EoM', the code has been developed by the author and his research group over a number of years. The code is published under an open source license, and is freely available online.

6.7 Example Systems

6.7.1 Rigid Rider Bicycle

As an example of a complex multibody system, the method above will be applied to a well known example vehicle described in the literature[5], the rigid rider bicycle. The bicycle is modelled as four rigid bodies: the frame and rider treated as one body, the handlebar and fork assembly, the front wheel, and the rear wheel, as shown in Figure 6.4. As the name implies, the rider of the bicycle is assumed to be rigidly fixed to the frame and any relative motion is not modelled. The fork assembly is attached to the frame with a revolute joint representing the steering head bearing. The location of the steering head bearing is not given, but it must lie on the steer axis, which is defined by the axis tilt, and the trail (the distance from the front wheel ground contact to the point where the steer axis intersects the ground plane, with positive trail indicating the intersection lies in front of the wheel contact). Each wheel is also attached with a revolute joint at its centre to model the wheel bearings, the front wheel to the fork assembly, and the rear wheel to the frame. Friction is neglected in all three joints. The wheels are assumed to be perfectly round, uniform, and 'knife edged', i.e., the width of the tire is ignored. The bottom of each wheel maintains contact with the road surface. The contact is treated as a rolling constraint, i.e., the lowest point on the wheel has zero velocity, and must be in the horizontal ground plane. The vertical and longitudinal motion of the contact point is treated as holonomic, i.e., the constraints are written in terms of position and orientation.

One interesting feature of the rigid rider model is that it is nonholonomic; there are no restrictions on the lateral location of contact points of the wheels with the road, but their velocities must always be zero at the contact point. The contact point may displace laterally, but only when the wheel steer history allows it. The model as presented in the literature has two degrees of freedom,

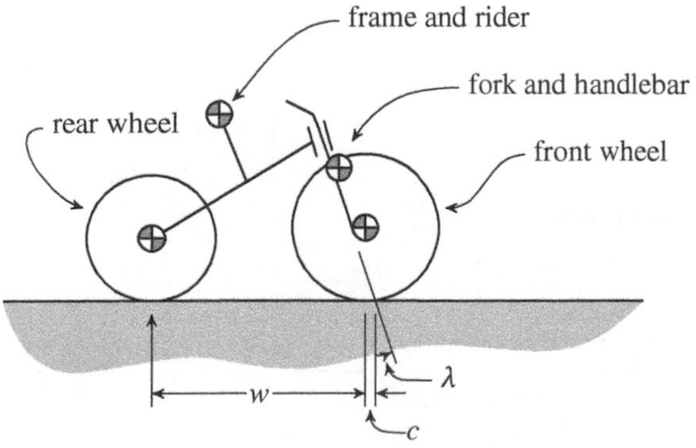

Figure 6.4 The rigid rider bicycle model is composed of four bodies: the frame and rider, the fork, and the front and rear wheels.

corresponding to the lean (roll) angle ϕ and steer angle δ motions. In the multibody model, additional lateral and yaw motions are allowed to occur by the constraints in the model, but appear as neutrally stable *rigid body modes*, i.e., the eigenvalue associated with these motions has a value of zero, meaning that there is no resulting response if the system has an initial displacement in this direction. The equations of motion are linearized around a fixed forward speed in the range from 0 to 10 m/s. Note that the sign convention used is the SAE standard, with positive z pointing downward, and that the rear wheel contact point is taken as the origin. The mass and inertia properties of all the components are included in the reference, along with the dimensions, which are reproduced in Table 6.1.

Table 6.1 Rigid rider bicycle parameter values

Quantity	Notation	Value
wheelbase	w	1.02 m
trail	c	0.08 m
steer axis tilt (π/2-head angle)	λ	π/10 rad (90°-72°)
rear wheel		
radius	r_{R}	0.3 m
mass	m_{R}	2 kg
moments of inertia	$(I_{\mathrm{R}xx}, I_{\mathrm{R}yy})$	(0.0603, 0.12) kg m^2
frame and rider		
centre of mass location	$(x_{\mathrm{B}}, z_{\mathrm{B}})$	(0.3, −0.9) m
mass	m_{B}	85 kg
inertia matrix	$\begin{bmatrix} I_{\mathrm{B}xx} & 0 & I_{\mathrm{B}xz} \\ 0 & I_{\mathrm{B}yy} & 0 \\ I_{\mathrm{B}xz} & 0 & I_{\mathrm{B}zz} \end{bmatrix}$	$\begin{bmatrix} 9.2 & 0 & 2.4 \\ 0 & 11 & 0 \\ 2.4 & 0 & 2.8 \end{bmatrix}$ kg m^2
handlebar and fork assembly		
centre of mass location	$(x_{\mathrm{H}}, z_{\mathrm{H}})$	(0.9, −0.7) m
mass	m_{H}	4 kg
inertia matrix	$\begin{bmatrix} I_{\mathrm{H}xx} & 0 & I_{\mathrm{H}xz} \\ 0 & I_{\mathrm{H}yy} & 0 \\ I_{\mathrm{H}xz} & 0 & I_{\mathrm{H}zz} \end{bmatrix}$	$\begin{bmatrix} 0.05892 & 0 & -0.00756 \\ 0 & 0.06 & 0 \\ -0.00756 & 0 & 0.00708 \end{bmatrix}$ kg m^2
front wheel		
radius	r_{F}	0.35 m
mass	m_{F}	3 kg
mass moments of inertia	$(I_{\mathrm{F}xx}, I_{\mathrm{F}yy})$	(0.1405, 0.28) kg m^2

Note: from Meijaard et al[5]

Matrices

The matrices generated in the equations of motion are presented below. The first four terms presented are the mass, damping, stiffness, and velocity matrices. All are square, and of size 24×24, as each of the four bodies requires six rows.

The mass matrix **M** is almost entirely diagonal. The first 6×6 block is the frame, the second block is the fork, the third is the front wheel, and the fourth is the rear wheel. All of the cross products of inertia are zero for the wheels, and only the xz entry is nonzero for the frame and the fork. The damping matrix **L** is composed of the sum of the terms that come from physical dampers, which in this case are zero, and the terms from the centripetal and gyroscopic effects. All four bodies display centripetal effects, as all have the same forward speed. When combined with translation in x, rotation around z requires force in y, and rotation around y requires force in z. The wheels also exhibit gyroscopic moments due to their rotation. When combined with rotation around the y axis, rotation around x requires a moment around z, and rotation around z requires a moment around x.

M = 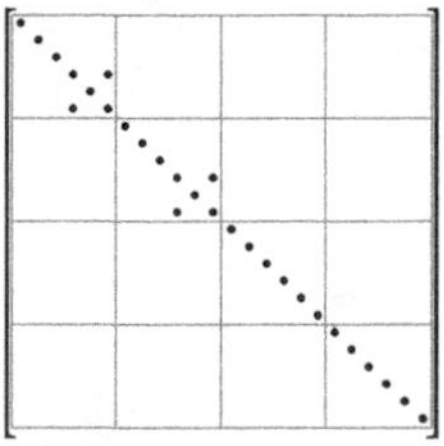 L = 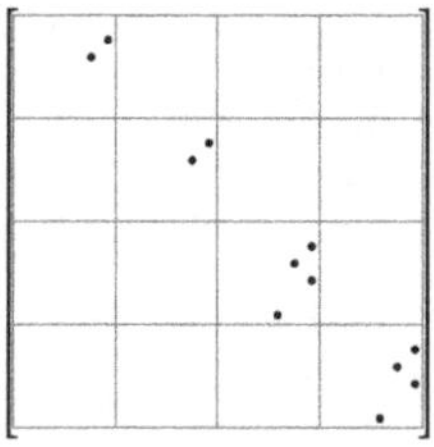

The stiffness matrix **K** is composed of the sum of the terms come from physical springs, which in this case are all zero, and the terms from the tangent stiffness effects. The tangent stiffness terms are moment terms (i.e., they are all in the bottom three rows of each block, and the result from the misalignment of the reference frames of each body and the forces or moments acting on them in the equilibrium condition.

The velocity matrix **V** results from the kinematics of the problem. All four bodies have the same forward speed in x, so a rotation around y results in a motion in z, and a rotation around z results in a motion in y.

K = 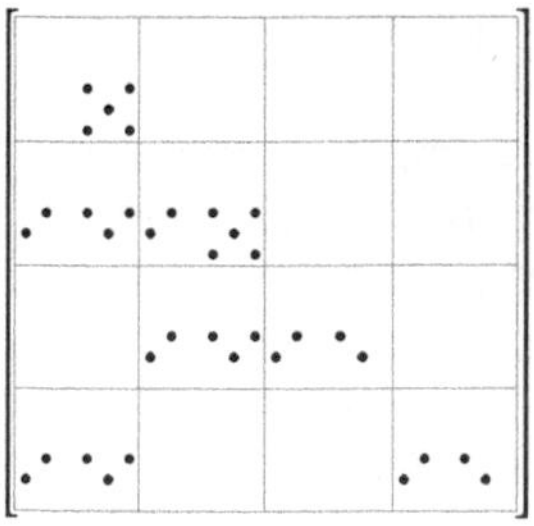 V = 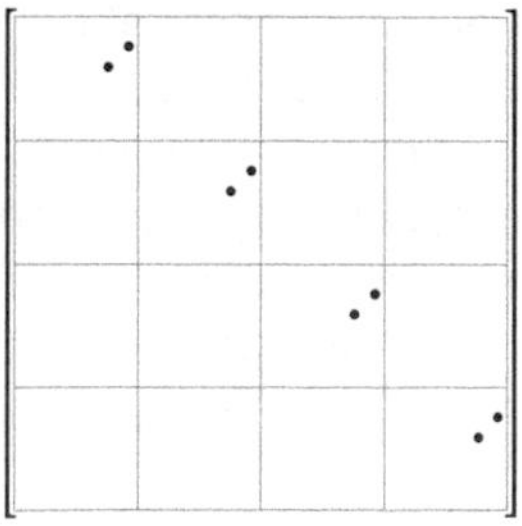

The final two matrices presented are the holonomic and nonholonomic constraint Jacobians. Neither matrix is square, but both have the same number of columns as the mass, damping, etc. The number of rows depends on the number of constraints present. The first five rows of the holonomic constraint matrix come from the steering head bearing; they couple the frame and fork motions, and so only appear in the columns corresponding to those bodies. The next five rows are the rear axle connecting the frame and rear wheel. Rows eleven through fifteen are the front wheel bearing connecting the fork and front wheel motions. Finally, there are two rows for each of the wheel contacts with the ground, and a constraint on the longitudinal motion of the frame. The final row could alternatively be moved to the nonholonomic constraint matrix to represent the fixed forward speed, but this would simply add another rigid body mode, i.e., a zero eigenvalue, to the result. The two rows in the nonholonomic constraint matrix represent the zero lateral speed of each of the tire contacts.

$J_h =$ 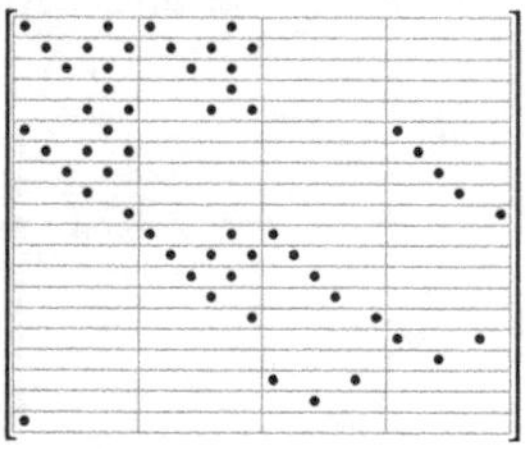

$J_{nh} =$ 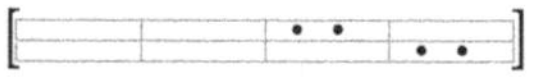

Results

Figure 6.5 shows an eigenvalue plot generated from the solution of the equations of motion, as a function of the forward speed. The plot is interesting as it shows four distinct regions with differing stability properties. At very low speed, below approximately 0.6 m/s, the bicycle has a relatively fast exponential instability, with a time constant less than 0.2 s. From 0.6 m/s, to approximately 4.3 m/s, the bicycle is still unstable, but with an oscillatory response. The oscillations are approximately 0.5 Hz. From 4.3 m/s to approximately 6.0 m/s, there is a region of oscillatory but stable behaviour. Above 6.0 m/s, another exponential instability appears, but with a very slow growth, with a time constant over 6.0 s.

The model becomes particularly interesting near the transition speed from unstable to stable, as the real part of the root crosses zero, the damping also goes to zero. The vibration metrics of the model at $u = 4.3$ m/s are given in Table 6.2. The damping ratio is very small for the oscillatory mode at that speed, meaning that the bike is quite susceptible to vibration at its natural frequency of approximately 5.5 Hz. The frequency response of the roll angle and the steer angle per unit steer torque is shown in Figure 6.6.

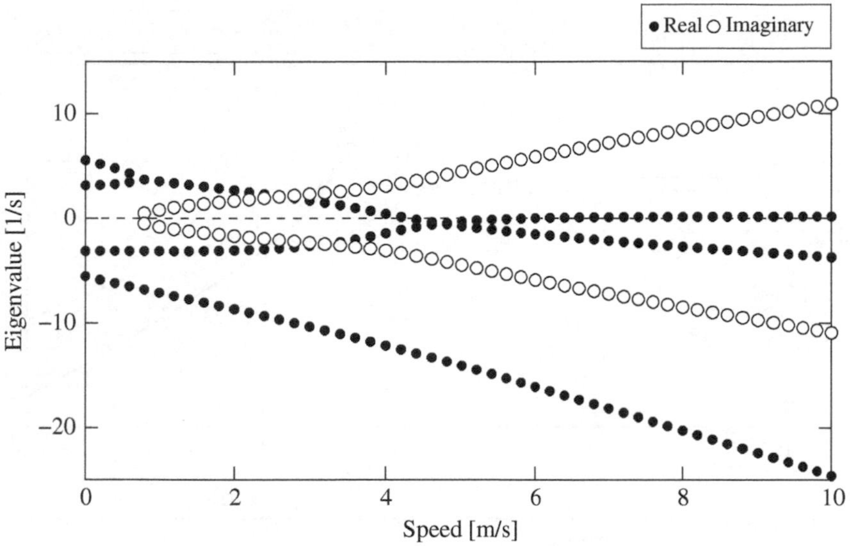

Figure 6.5 The eigenvalues of the rigid rider bicycle model show four different regions of stability. The oscillatory mode transitions from unstable to stable at a forward speed of approximately $u = 4.3$ m/s.

Table 6.2 Rigid rider bicycle model vibration metrics

No.	ω_n [Hz]	ζ	τ [s]	λ [s]
1	–	–	7.8592×10^{-2}	–
2	5.4834×10^{-1}	2.9594×10^{-3}	9.8076×10^{1}	1.8237×10^{0}
3	–	–	1.0263×10^{0}	–

Note: table includes natural frequency (ω_n), damping ratio (ζ), time constants (τ), and wavelength (λ) when forward speed is set at $u = 4.3$ m/s, discarding the rigid body modes

6.7.2 Multibody Quarter Car

The second example is an enhanced quarter car model, conceptually similar to the one described in Chapter 4, but much more elaborate[1]. It uses a four bar style linkage to model the suspension mechanism. The model consists of four rigid bodies: the chassis, the upper and lower A-arm suspension links, and a lumped wheel, hub, and upright. The system is shown in Figure 6.7. The properties of the bodies are given in Tables 6.3 and 6.4. The mechanism is assumed to be in its equilibrium position under the action of gravity, and the preloads are

1 Courtesy S Els et al in the Vehicle Dynamics Group at the University of Pretoria.

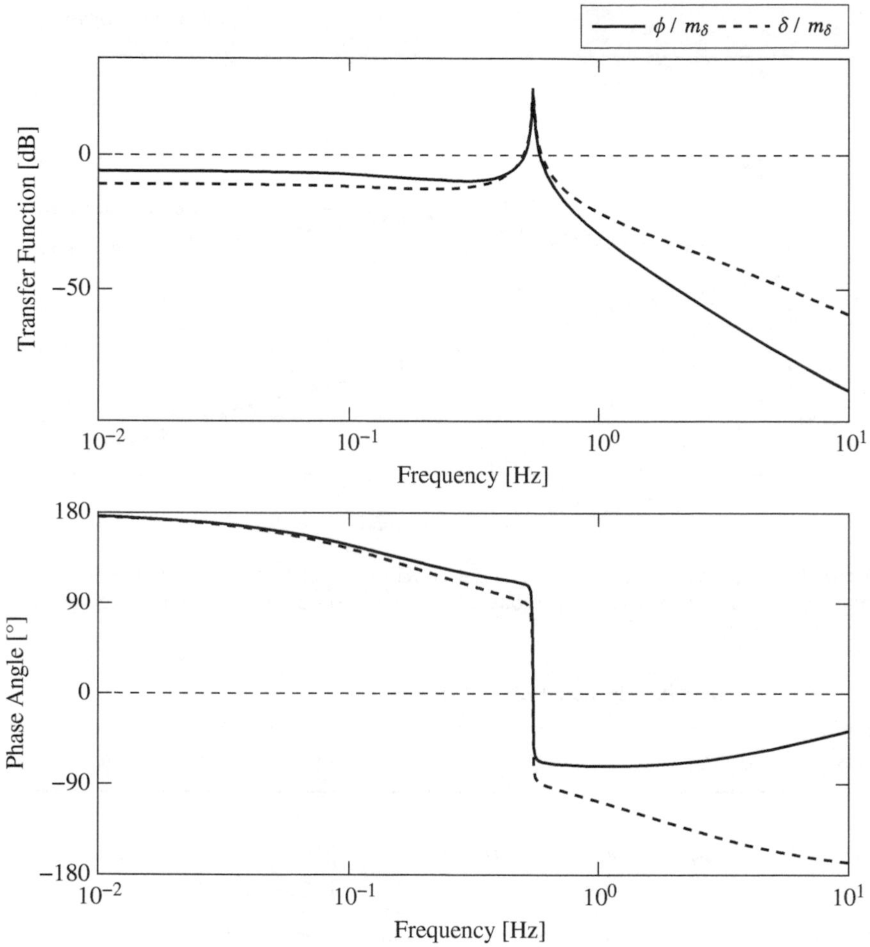

Figure 6.6 The roll angle ϕ and steer angle δ of the rigid rider bicycle in response to sinusoidal steer moment m_δ at a forward speed of $u = 4.3$ m/s. The resonance at approximately 0.55 Hz is clearly visible.

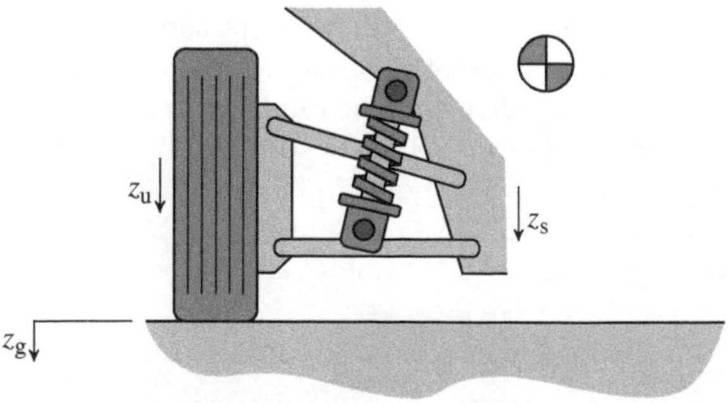

Figure 6.7 The multibody quarter car model is composed of four bodies: the chassis, the wheel+hub, and the upper and lower suspension arms.

Table 6.3 Multibody quarter car body locations and masses

Body name	Location [m] (rx, ry, rz)			Mass [kg]
Chassis	0.000,	0.000,	0.300	75
Wheel	0.000,	0.600,	0.250	15
Lower A-arm	0.000,	0.325,	0.750	1
Upper A-arm	0.000,	0.305,	0.400	1

Table 6.4 Multibody quarter car body inertia properties

Body name	Inertia [kg m^2] (I_{xx}, I_{yy}, I_{zz}; I_{xy}, I_{yz}, I_{zx})					
Chassis	0.000	0.000,	0.000;	0.000,	0.000,	0.000
Wheel	0.234,	0.234,	0.469;	0.000,	0.000,	0.000
Lower A-arm	0.010,	0.000,	0.010;	0.000,	−0.001,	0.000
Upper A-arm	0.008,	0.000,	0.008;	0.000,	0.000,	0.000

computed based on this assumption. The chassis is constrained to allow vertical translation only. The upper and lower A-arms are each fixed to the chassis with revolute joints. The lower arm is also fixed to the wheel carrier with a revolute joint. The upper arm connection uses a modified hinge that only restricts motion in the vertical and lateral directions, with no restriction on rotations or axial translation; this is to prevent the mechanism from being overconstrained. The properties of the connections are given in Tables 6.5. The suspension spring

Table 6.5 Multibody quarter car connection locations and directions

Connection name	Location [m] (r_x, r_y, r_z)			Unit Axis (u_x, u_y, u_z)		
Lower ball joint	0.000,	0.500,	0.050	1.000,	0.000,	0.000
Lower A-arm pivot	0.000,	0.150,	0.100	1.000,	0.000,	0.000
Upper ball joint	0.000,	0.460,	0.400	1.000,	0.000,	0.000
Upper A-arm pivot	0.000,	0.150,	0.400	1.000,	0.000,	0.000
Chassis slider	0.000,	0.000,	0.450	0.000,	0.000,	1.000
Tire, vertical	0.000,	0.600,	0.000	0.000,	0.000,	1.000
Suspension spring	0.000,	0.150,	0.570	–	–	–
–	0.000,	0.430,	0.060	–	–	–

Table 6.6 Multibody quarter car body stiffness and damping

Connection name	Stiffness [N/m]	Damping [Ns/m]
Tire, vertical	40 000	0
Suspension spring	6 100	1 100

and damper are linear and coaxial, acting between the lower control arm and the chassis. The tire was modelled as a linear bushing between the wheel and the ground, with a unidirectional vertical stiffness and damping. The properties of the flexible connectors are given in Table 6.6.

Matrices

The matrices from the multibody quarter car model are listed below. The mass matrix is again nearly diagonal. The first block corresponds to the chassis; the inertia values are neglected as the chassis is constrained against rotation, so they have no effect. The wheel/hub/upright is treated as a cylinder for the purpose of computing the mass moments of inertia. The upper and lower arms are treated as slender rods, and their inertia matrices are computed accordingly. The stiffness matrix is quite complex, with many nonzero entries. It is computed as a sum of three matrices, each of which are presented individually.

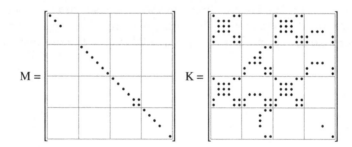

M =

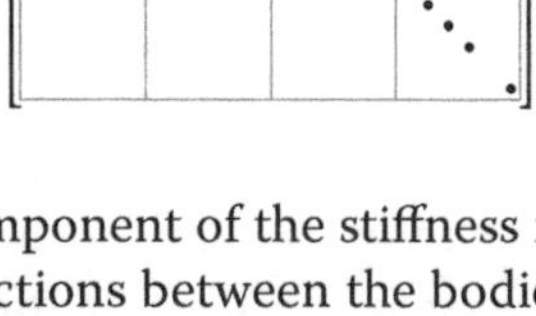

K =

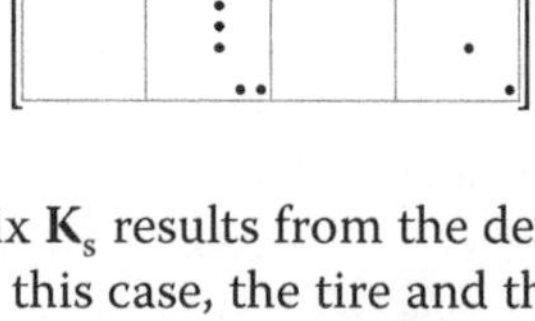

The first component of the stiffness matrix $\mathbf{K}_s$ results from the deflection of flexible connections between the bodies, in this case, the tire and the suspension spring. The four small 3×3 blocks all result from the suspension spring, and the individual point comes from the tire vertical stiffness. The damping matrix, not shown, is of identical form the to this matrix, except it lacks the term in the location of the tire stiffness, as the tire is treated as undamped. The tangent stiffness matrix $\mathbf{K}_t$ is much more complex, and results from the many preload forces in each of the constraint elements, and in the tire and suspension spring.

$K_s =$ 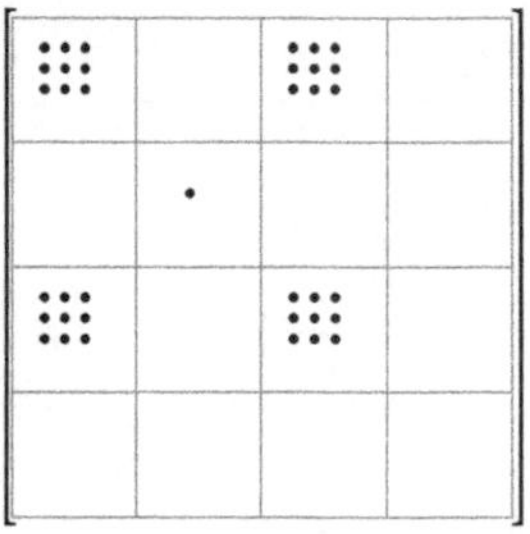 $K_t =$ 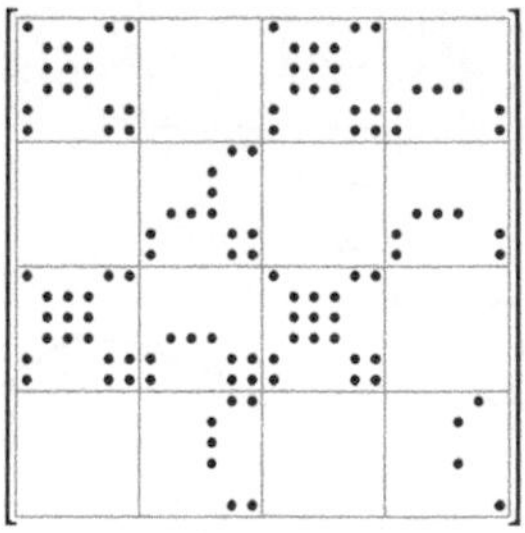

The last component of the stiffness matrix comes only from the effect of the weight force acting on each of the bodies. The weight force in the z axis results in coupling terms between the x and y axes, as in the bicycle example. Finally, the constraint matrix is included. There are no nonholonomic constraints in the model. The form clearly shows the constraints that connect the upright to the lower arm (five rows), the chassis to the lower arm (five rows), the upright to the upper arm (two rows), and the chassis to the upper arm (five rows). The last block of rows restrict the motion of the chassis to vertical translation only.

$K_l =$ 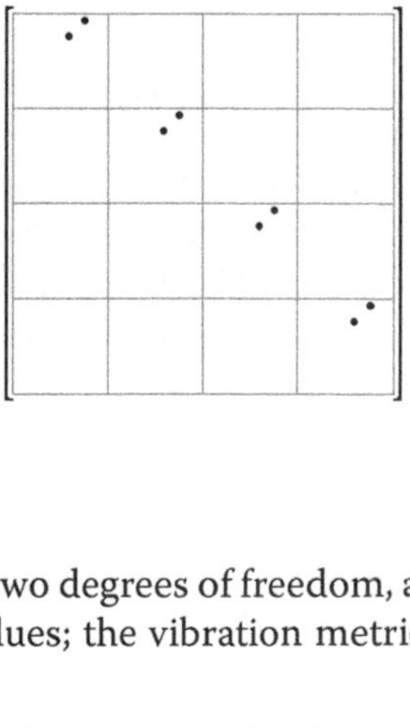 $J_h =$ 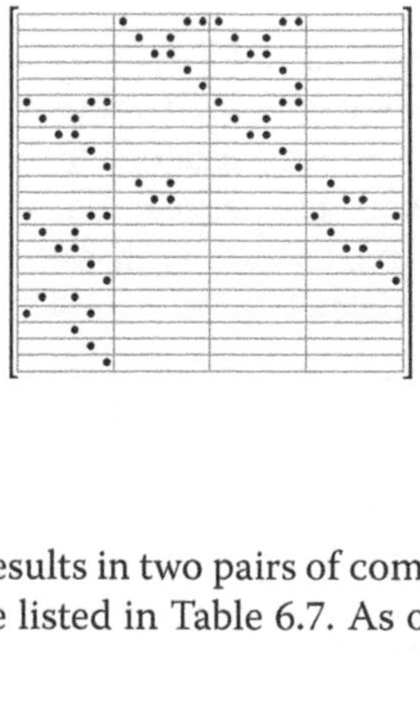

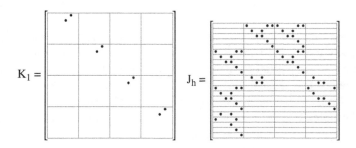

Results

The system is two degrees of freedom, and results in two pairs of complex conjugate eigenvalues; the vibration metrics are listed in Table 6.7. As one might

Table 6.7 Multibody quarter car vibration metrics

No.	ω_n [Hz]	ζ	τ [s]	λ [s]
1	8.1268×10^{0}	2.7170×10^{-1}	7.2080×10^{-2}	1.2786×10^{-1}
2	1.0052×10^{0}	3.9927×10^{-1}	3.9655×10^{-1}	1.0851×10^{0}

Note: table includes natural frequencies (ω_n), damping ratios (ζ), time constants (τ), and wavelengths (λ)

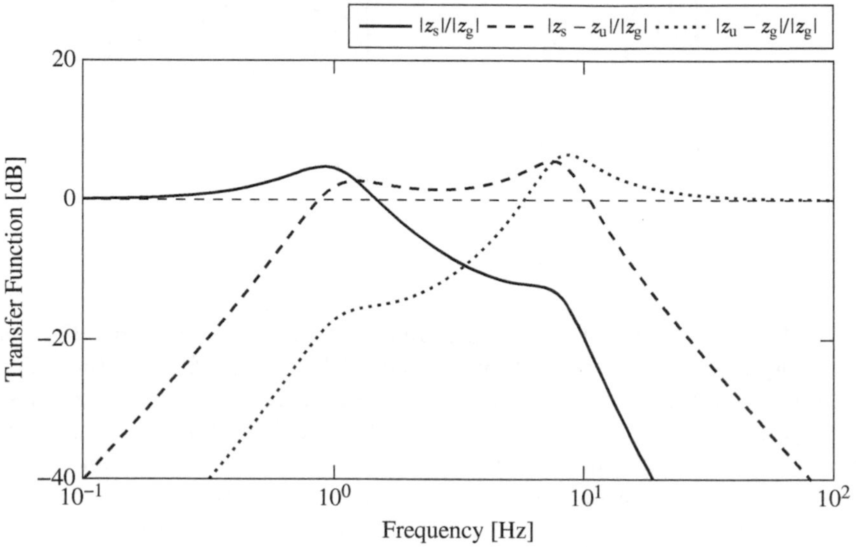

Figure 6.8 Despite the added complexity, the frequency response of multibody quarter car model is very similar to the lumped mass model.

expect, the high frequency oscillatory wheel hop mode appears, followed by the low frequency body motion. The forced response of the model is shown in Figure 6.8. An actuator under the wheel drives the motion, and three sensors measure the vertical motion of the sprung mass, the vertical motion of the wheel relative to the chassis, and the vertical tire compression. Despite the added complexity of the model, the appearance of the frequency response is clearly very similar to the classic lumped mass quarter car model presented in Figure 4.23. The low frequency input is absorbed by motion of the chassis, the midrange by the suspension travel, and the high frequency by the tire compression.

References

1 Amirouche, F., 2007. *Fundamentals of Multibody Dynamics: Theory and Applications.* Springer Science & Business Media.

2 Nikravesh, P.E., 1988. *Computer-aided Analysis of Mechanical Systems.* Prentice-Hall, Inc.

3 Shabana, A.A., 2013. *Dynamics of Multibody Systems.* Cambridge University Press.

4 Schiehlen, Werner, ed. *Multibody Systems Handbook*. Vol. 6. Berlin etc.: Springer.

5 Meijaard, J.P., Papadopoulos, J.M., Ruina, A., and Schwab, A.L., 2007. Linearized dynamics equations for the balance and steer of a bicycle: a benchmark and review. *Proceedings of the Royal Society of London A: Mathematical, Physical and Engineering Sciences* 463(2084), pp. 1955–1982.

7

Mathematics

The study of vehicle dynamics depends heavily on a foundation in mathematics. The discussions in this text will frequently rely on the equations of motion, and the associated analysis of these equations. It is expected that the reader will have some background, but this chapter will provide an overview of the more important topics. The coverage will be brief, but there are many other sources available to those readers who are interested in more detail.

Regarding the relevant equations, there are two primary classifications that will be of concern. These are the distinction between *algebraic* and *differential* equations, and the *linearity* of those equations.

7.1 Algebraic Equations

In the context of this chapter, algebraic equations are defined as any set of mathematical relationships between a set of variables, with the restriction that no rate dependence appears. In this setting, the equations are assumed to describe a dynamic system, so there is one independent variable, this being time, denoted t, and many dependent variables, gathered into a vector $\boldsymbol{x}$. While there may be functions of time present, derivatives or integrals with respect to time do not appear. Usually, algebraic equations will occur as sets or systems, i.e., there will be many variables and many coupled equations. Algebraic equations are sometimes defined more precisely as including only polynomials, but here the distinction is ignored, and no such restriction is implied. Algebraic equations are written as:

$$\boldsymbol{f}(\boldsymbol{x}, t) = \boldsymbol{0} \tag{7.1}$$

7.1.1 Nonlinear Algebraic Equations

Unless certain conditions are satisfied, algebraic equations are generally nonlinear. If any products of dependent variables, or any dependent variables raised

Fundamentals of Vehicle Dynamics and Modelling: A Textbook for Engineers with Illustrations and Examples, First Edition. Bruce P. Minaker.

Companion website: www.wiley.com/go/minaker/vehicle-dynamics

to any power, or any trigonometric, exponential, or other similar functions of the dependent variables appear, the equations are nonlinear. An example of a nonlinear set of algebraic equations is:

$$\begin{Bmatrix} 3\sqrt{x_1} + 4x_2 \\ 4x_1^2 + 3x_3 + t \\ x_1x_3 - 1 \end{Bmatrix} = \begin{Bmatrix} 0 \\ 0 \\ 0 \end{Bmatrix} \tag{7.2}$$

In the example, all of the individual equations are nonlinear; however, a single nonlinear equation is sufficient to for the entire system is considered to be nonlinear.

Before seeking a solution to a system or even a single nonlinear algebraic equation, one should first ask if any solution exists, or if more than one solution exists. Nonlinear algebraic equations pose a challenge in that there is no guarantee of a solution, or there may be multiple solutions. In fact, the concerns of *existence* and *uniqueness* of the solution are of prime importance in a purely mathematical approach to the problem. Engineers will often take a pragmatic approach, and overlook these issues. For example, if the equations accurately describe the kinematics of some type of physical mechanism, the existence of a solution is not usually in question. However, in many mechanisms, the problem of multiple solutions can sometimes appear. A common example is the well known four bar mechanism, where the equations will provide the expected solution, but may also allow another perhaps unexpected solution where the bars are crossed. In some cases, it may be possible to find a closed form algebraic solution to a set of nonlinear equations, but this is rarely possible for anything other than small and simple systems. Typically a numerical approach must be used, usually one involving recursion, such as Newton–Raphson.

7.1.2 Linear Algebraic Equations

In some cases, a system of algebraic equations can be classified as linear. If so, it can be written in a form where the relationships between the dependent variables appear only as a product with a matrix function of the independent variable, as so:

$$\mathbf{A}(t)\boldsymbol{x}(t) - \boldsymbol{f}(t) = \mathbf{0} \tag{7.3}$$

Note that the matrix $\mathbf{A}$ is not necessarily a function of time; in fact, there are many problems of interest where it is a constant. Dependence on time does not disqualify a system as linear. An example of a linear set of algebraic equations is:

$$\begin{bmatrix} 3 & 4 & 0 \\ 4 & 0 & 3 \\ 1 & 0 & 1 \end{bmatrix} \begin{Bmatrix} x_1 \\ x_2 \\ x_3 \end{Bmatrix} = \begin{Bmatrix} t + 3 \\ t \\ 0 \end{Bmatrix} \tag{7.4}$$

From a numerical solution viewpoint, linear algebraic equations are generally favoured over nonlinear, as the solution process is more predictable and straightforward. A number of methods have been developed to solve linear algebraic equations. The most popular approaches are based on Gaussian elimination, or lower-upper (LU) decomposition, where the matrix **A** is written as the product of a lower triangular matrix and an upper triangular matrix. Unlike the case with the recursive methods used on nonlinear equations, both of these methods will produce a solution in a predetermined number of operations. Unfortunately, the number of operations required to find the solution grows very rapidly with an increase in the number of equations. As such, usually the sheer volume of calculations required dictates that a computer solver will be the pragmatic choice for systems with more than five or six equations. In the event of very large systems, e.g. tens of thousands of equations or more, there are also methods based on recursion, such as Gauss–Seidel, that often deliver an equally accurate solution, but with fewer operations.

7.2 Differential Equations

A differential equation is one in which the rate of change of some value is related to the value itself. Differential equations are further classified as *partial* differential equations (PDEs), or *ordinary* differential equations (ODEs). In a partial differential equation, there are multiple independent variables; frequently these are the (x, y, z) coordinates in physical space. In an ordinary differential equation, there is only one independent variable, which is often time. Both may have many dependent variables. Partial differential equations, which are generally significantly more challenging to solve, are outside the scope of this text and will not be discussed here. Ordinary differential equations, however, are central, and will be further explained. The most general form of a set of ODEs, sometimes referred to as the *implicit form*, is written as:

$$\boldsymbol{f}(\boldsymbol{x}(t), \dot{\boldsymbol{x}}(t), t) = \mathbf{0} \tag{7.5}$$

Here the function $\boldsymbol{f}$ may be such that the $\dot{\boldsymbol{x}}$ terms cannot be isolated from the rest of the equation. However, implicit ODEs are not particularly common when dealing with problems involving the motion of mechanical systems, and usually, they will be expressed in standard or *explicit first order form*:

$$\dot{\boldsymbol{x}}(t) = \boldsymbol{f}(\boldsymbol{x}, t) \tag{7.6}$$

The function $\boldsymbol{f}(\boldsymbol{x}, t)$ is often referred to as the *forcing function*. The solution to a differential equation contains two components, the *homogeneous* or unforced solution, and the *particular* or forced solution of the ODE.

$$\boldsymbol{x}(t) = \boldsymbol{x}_\mathrm{h}(t) + \boldsymbol{x}_\mathrm{p}(t) \tag{7.7}$$

In a mechanical vibration or motion problem, the unforced solution is the motion that occurs due only to the initial state of the system, and not as a response to any external influence. It is found by solving the differential equation with the forcing function set to zero. It is important to note that even in the case of a system with a nonzero forcing function, the unforced component of the solution will still be present. This often confusing result can be reconciled by recognizing that adding the unforced solution to the forced solution is equivalent to adding a zero to the forcing function, i.e., the addition of the unforced solution still satisfies the differential equation. The unforced solution is often referred to as the *transient* solution, as it typically dissipates with time.

In order to find the solution to a differential equation, a known solution at a specific point in time must also be provided. Typically, this solution satisfies the equations at the initial time, and so is referred to as the *initial conditions.*

7.2.1 Nonlinear Differential Equations

Like algebraic equations, differential equations are also generally nonlinear, unless certain conditions apply. An example of a system of nonlinear differential equations, with initial conditions, is:

$$\begin{Bmatrix} \dot{x}_1 \\ \dot{x}_2 \\ \dot{x}_3 \end{Bmatrix} = \begin{Bmatrix} 3\sqrt{x_1} + 4x_2 \\ 4x_1^2 + 3x_3 + t \\ x_1 x_3 - 1 \end{Bmatrix}, \quad \begin{Bmatrix} x_1(0) \\ x_2(0) \\ x_3(0) \end{Bmatrix} = \begin{Bmatrix} 1 \\ 0 \\ 1 \end{Bmatrix} \tag{7.8}$$

As is the case with algebraic equations, sets of nonlinear ODEs will typically be complicated enough that a numerical method is the only practical means of obtaining a solution. It is important to note that even small changes to the initial conditions can have a dramatic effect on the character of the solution of a nonlinear ODE.

7.2.2 Linear First Order ODEs

Throughout this text, the reader will frequently encounter linear or linearized ODEs. These linear ODEs will often provide a less accurate description of the problem than their nonlinear counterparts, but they remain popular. One reason for this is that they often allow a model that is still sufficiently accurate to capture the relevant phenomena, but they offer more opportunity in terms of solution technique. In fact, the additional information that is provided by a linear set of equations will often make it worthwhile to linearize an existing set of nonlinear equations. Unless the phenomena of interest is highly nonlinear, a strong argument can often be made for restricting the analysis to the linear region. A set of linear ODEs can be expressed as shown:

$$\dot{\boldsymbol{x}}(t) = \mathbf{A}(t)\boldsymbol{x}(t) + \boldsymbol{f}(t) \tag{7.9}$$

The matrix **A** may potentially be a function of time, but frequently it will be a constant. In this case, the problem may be referred to as *linear time invariant* (LTI). The primary distinction of LTI problems is that the questions of existence and uniqueness are already answered: there will be one solution only, and the form is known. Consider the unforced first order LTI ODE:

$$\dot{\boldsymbol{x}}(t) = \mathbf{A}\boldsymbol{x}(t) \tag{7.10}$$

The form of the solution is assumed to be exponential, with all components of the vector sharing the same time history behaviour, but with each component having a unique magnitude, i.e., the time varying vector $\boldsymbol{x}(t)$ is assumed to be the product of a constant vector $\boldsymbol{x}_0$ and a time varying scalar e^{st}.

$$\boldsymbol{x}(t) = \boldsymbol{x}_0 \mathrm{e}^{st} \tag{7.11}$$

Note that when $t = 0$, $\boldsymbol{x}(t) = \boldsymbol{x}_0$. Differentiating and substituting into the original equation gives:

$$\boldsymbol{x}_0 s\mathrm{e}^{st} = \mathbf{A}\boldsymbol{x}_0 \mathrm{e}^{st} \tag{7.12}$$

or:

$$[\mathbf{I}s - \mathbf{A}]\boldsymbol{x}_0 \mathrm{e}^{st} = \mathbf{0} \tag{7.13}$$

If one finds the values of s such that:

$$[\mathbf{I}s - \mathbf{A}]\boldsymbol{x}_0 = \mathbf{0} \tag{7.14}$$

then Equation (7.13) will always be satisfied, assuming finite values of the exponential term, and Equation (7.11) then solves the ODE. Equation (7.14) is referred to as the *eigenvector* problem.

7.2.3 Eigen Analysis

The most common tool for solution of linear ODEs is eigen analysis. The word 'eigen' is German, and translated to English means 'own', or 'characteristic'. Eigen analysis consists of two related problems: the previously mentioned eigenvector problem, and the associated *eigenvalue* problem.

The *dimension* of the problem, i.e., the number of rows and columns of the matrix $[\mathbf{I}s - \mathbf{A}]$, is denoted as n. When the matrix $[\mathbf{I}s - \mathbf{A}]$ is multiplied by the vector $\boldsymbol{x}_0$, each of the n rows in turn is multiplied, or equivalently, the dot product of each row with $\boldsymbol{x}_0$ is found. Recall that if two vectors have a dot product of zero, they must be perpendicular. When one allows that the vector $\boldsymbol{x}_0$ is nonzero, then all n rows of $[\mathbf{I}s - \mathbf{A}]$ must be perpendicular to $\boldsymbol{x}_0$ in order to satisfy Equation (7.14). However, at most $n-1$ vectors of dimension n can be mutually perpendicular to each other, while all remaining perpendicular to $\boldsymbol{x}_0$. This means that one of the rows of $[\mathbf{I}s - \mathbf{A}]$ must be linear combination of the others, and therefore the matrix must be singular, i.e., it has a determinant of zero. The eigenvalue problem is formally posed as:

$$p(s) = \det[\mathbf{I}s - \mathbf{A}] = 0 \tag{7.15}$$

The eigenvalue problem seeks the set of values s that force the matrix $[\mathbf{I}s - \mathbf{A}]$ to be singular. The equation formed by expanding the determinant, $p(s)$, a polynomial of degree n in s, and setting it to zero, is called the *characteristic equation.*

$$s^n + p_{n-1}s^{n-1} + \ldots + p_1 s + p_0 = 0 \tag{7.16}$$

There are n roots of the characteristic equation; these are the eigenvalues. Each eigenvalue in turn is used to find a singular matrix, and the eigenvector is then perpendicular to all the rows of that singular matrix. More formally, the eigenvector lies in the *null space* of $[\mathbf{I}s - \mathbf{A}]$. As a result, there are also n eigenvectors, and each eigenvector is associated with a particular eigenvalue. It is the direction only of the eigenvector that is important to satisfy Equation (7.14). The eigenvector is of arbitrary magnitude and so the solution is not unique. For consistency of the solution, the eigenvectors are usually normalized to a unit length, but this is not required.

The interest in the eigenvector problem is at least partially due to the fact that the unforced solution of an LTI ODE can be found directly from this type of analysis, although it does have other uses as well. Because each eigenvalue/eigenvector pair is a solution to Equation (7.14), the homogeneous solution to the LTI problem is a linear combination of the contribution of each pair.

$$\boldsymbol{x}(t) = c_1\boldsymbol{x}_1 e^{s_1 t} + c_2\boldsymbol{x}_2 e^{s_2 t} + \ldots + c_n\boldsymbol{x}_n e^{s_n t} \tag{7.17}$$

The individual weights of each eigenvector depend on the initial conditions of the problem. There are two important consequences of the form of the solution. First, the unforced solution of an LTI ODE is linear in the initial conditions, i.e., if a given set of initial conditions is scaled by some value, the unforced solution scales by the same value. Secondly, the characteristic behaviour is independent of the initial conditions, i.e., changing the initial conditions does not change the eigenvalues.

It is often the case that the eigenvalues and eigenvectors contain complex numbers. Complex eigenvalues always come in pairs; any time a complex eigenvalue appears, its complex conjugate must also be present. The presence of complex roots simply indicates that the solution contains oscillatory terms in

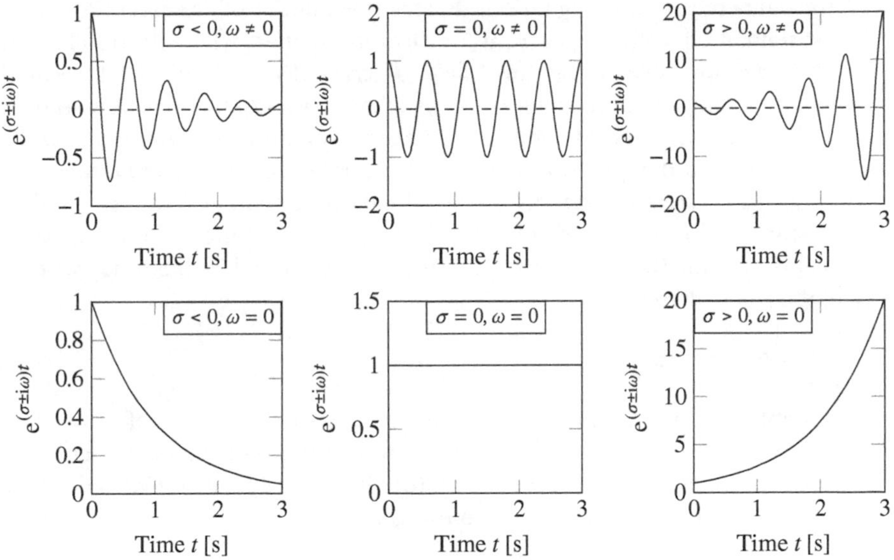

Figure 7.1 The time history of the solution to a linear first order system can be either exponential or sinusoidal, and either stable, neutrally stable, or unstable.

addition to the exponential terms, and is shown through Euler's formula:

$$e^{st} = e^{(\sigma+i\omega)t} = e^{\sigma t}(\cos(\omega t) + i\sin(\omega t)) \tag{7.18}$$

The requirement that the state vector $\boldsymbol{x}$ be real is met by the fact that the eigenvectors that are associated with a complex conjugate pair of eigenvalues will also be complex conjugates. Careful expansion of a pair of complex conjugate terms shows that the imaginary terms cancel and the result is real only.

$$(a + bi)e^{(\sigma+i\omega)t} + (a - bi)e^{(\sigma-i\omega)t} = 2e^{\sigma t}(a\cos(\omega t) - b\sin(\omega t)) \tag{7.19}$$

An examination of the form of the result shows that the real part of a complex root determines the rate of growth or decay of the amplitude of the solution, while the imaginary part determines the frequency of oscillation. The result is illustrated in Figure 7.1.

The real and imaginary parts of the root may instead be represented as the *natural frequency* ω_n and the *damping ratio* ζ, where:

$$\omega_n = \sqrt{\sigma^2 + \omega^2}, \quad \zeta = \frac{-\sigma}{\sqrt{\sigma^2 + \omega^2}} \tag{7.20}$$

Conversely:

$$\omega = \sqrt{1 - \zeta^2}\omega_n = \omega_d, \quad \sigma = -\zeta\omega_n \tag{7.21}$$

Here, the term ω_d is known as the *damped natural frequency*.

The value of the damping ratio indicates the amount of damping present in the system. It typically ranges from $\zeta = 0$, or undamped, where the oscillations in the solution never decay, to $\zeta = 1$, or *critically damped*, the threshold where the solution transitions from oscillatory to exponential. A damping ratio between zero and one indicates that the system is *underdamped*, while a system with a damping ratio above one is referred to as *overdamped*. An overdamped system will not display oscillatory motion in its free solution (but may still do so in its forced solution). A negative damping ratio indicates an unstable system. The rate of decay is often characterized by the *time constant* τ of the root, where:

$$\tau = -\frac{1}{\sigma} = \frac{1}{\zeta\omega_\mathrm{n}} \tag{7.22}$$

In a stable system, the amplitude shrinks to approximately 37% of its original size in each time constant ($\mathrm{e}^{-1} \approx 0.37$); after four time constants, it is less than 2% of its original size. The *wavelength* λ (or alternatively the *period*) is the time required for one oscillation, and is found as:

$$\lambda = \frac{2\pi}{\omega} \tag{7.23}$$

The number of oscillations that will occur before the amplitude shrinks to almost zero can be expressed as a function of the damping ratio only:

$$\frac{4\tau}{\lambda} = \frac{4\omega\tau}{2\pi} = \frac{2}{\pi}\frac{\sqrt{1-\zeta^2}\omega_\mathrm{n}}{\zeta\omega_\mathrm{n}} = \frac{2}{\pi}\frac{\sqrt{1-\zeta^2}}{\zeta} \tag{7.24}$$

As can be seen from the expression, even underdamped systems may not show significant oscillation. Even a damping value of $\zeta = 1/\sqrt{2} \approx 0.707$ is too large for one full cycle to appear before the amplitude crosses the 2% threshold.

Sometimes, the special case of *repeated roots* will occur, when a particular eigenvalue appears more than once in the solution. This is the case if the damping is exactly critical, or alternatively, some frequently encountered problems can result in multiple eigenvalues of zero. More precisely, the eigenvalues are termed *degenerate*, and the number of unique eigenvalues is less than the degree of the characteristic polynomial. The presence of repeated roots has two effects. First, the form of the solution of the associated LTI ODE will change. The terms associated with the repeated root will be multiplied by increasing powers of t. For example, in the case where s_1 appears twice:

$$\boldsymbol{x}(t) = c_1\boldsymbol{x}_1\mathrm{e}^{s_1 t} + c_2\boldsymbol{x}_2 t\mathrm{e}^{s_1 t} + \ldots + c_n\boldsymbol{x}_n\mathrm{e}^{s_n t} \tag{7.25}$$

Additionally, the homogeneous solution of the eigenvector problem becomes more complicated. Again, in the case where s_1 appears twice, the dimension of the null space of $[\mathbf{A} - s_1\mathbf{I}]$ will grow to two, and the repeated eigenvalue will be associated with any two orthogonal eigenvectors that are in the null space.

This implies that in the case of repeated roots, even the normalized eigenvectors are not unique.

If the **A** matrix of the system of interest is 2×2, the characteristic equation is quadratic and may be expanded and the roots solved directly. In practice, however, most problems of interest are much larger, and the corresponding characteristic equation is of a much higher degree, where no closed form solutions exist. In this case, the eigenvalue problem must be solved by an iterative numerical method, e.g., the QR algorithm or a variant, where the **A** matrix is decomposed into the product of an orthogonal matrix and an upper triangular matrix.

7.2.4 Linear Second Order ODEs

Usually, higher order differential equations will be reduced to first order to allow an eigenvalue analysis. Any nth order differential equation can be equivalently represented as n first order equations. However, because of their frequent appearance in mechanical vibration problems, the second order linear ODE problem is also of special interest. Consider the second order system:

$$\mathbf{M}\ddot{\boldsymbol{z}}(t) + \mathbf{L}\dot{\boldsymbol{z}}(t) + \mathbf{K}\boldsymbol{z}(t) = \mathbf{0} \tag{7.26}$$

Here, **M** is referred to as the *mass matrix*, **L** is the *damping matrix*, and **K** is the *stiffness matrix*. The second order equation can be written as a first order equation by first solving for $\ddot{\boldsymbol{z}}$.

$$\ddot{\boldsymbol{z}} = -\mathbf{M}^{-1}\mathbf{K}\boldsymbol{z} - \mathbf{M}^{-1}\mathbf{L}\dot{\boldsymbol{z}} \tag{7.27}$$

The first order form of the equations is double in size of the second order form.

$$\begin{Bmatrix} \dot{\boldsymbol{z}} \\ \ddot{\boldsymbol{z}} \end{Bmatrix} = \begin{bmatrix} \mathbf{0} & \mathbf{I} \\ -\mathbf{M}^{-1}\mathbf{K} & -\mathbf{M}^{-1}\mathbf{L} \end{bmatrix} \begin{Bmatrix} \boldsymbol{z} \\ \dot{\boldsymbol{z}} \end{Bmatrix} \tag{7.28}$$

If the vector $\boldsymbol{x}(t)$ is defined:

$$\boldsymbol{x}(t) = \begin{Bmatrix} \boldsymbol{z}(t) \\ \dot{\boldsymbol{z}}(t) \end{Bmatrix} \tag{7.29}$$

then the problem is now familiar.

$$\dot{\boldsymbol{x}}(t) = \mathbf{A}\boldsymbol{x}(t) \tag{7.30}$$

Consider a single second order system with two eigenvalues:

$$m\ddot{z} + c\dot{z} + kz = 0 \tag{7.31}$$

Reducing to first order gives:

$$\begin{Bmatrix} \dot{z} \\ \ddot{z} \end{Bmatrix} = \begin{bmatrix} 0 & 1 \\ -k/m & -c/m \end{bmatrix} \begin{Bmatrix} z \\ \dot{z} \end{Bmatrix} \tag{7.32}$$

The characteristic equation is:

$$p(s) = \det\left[\begin{bmatrix} s & 0 \\ 0 & s \end{bmatrix} - \begin{bmatrix} 0 & 1 \\ -k/m & -c/m \end{bmatrix}\right]$$
$$= \det\left[\begin{bmatrix} s & -1 \\ k/m & s + c/m \end{bmatrix}\right]$$
$$= s^2 + (c/m)s + k/m = 0 \tag{7.33}$$

In this case, the characteristic equation is a quadratic, and the roots can be found directly.

$$s = \frac{-\frac{c}{m} \pm \sqrt{\left(\frac{c}{m}\right)^2 - \frac{4k}{m}}}{2} \tag{7.34}$$

In order to have an oscillatory case, the roots must be complex, such that:

$$s = -\frac{c}{2m} \pm \sqrt{\left(\frac{c}{2m}\right)^2 - \frac{k}{m}} = -\frac{c}{2m} \pm \mathrm{i}\sqrt{\frac{k}{m} - \left(\frac{c}{2m}\right)^2} \tag{7.35}$$

Recognizing the real and imaginary parts of the root:

$$\omega_\mathrm{n} = \sqrt{\sigma^2 + \omega^2} = \sqrt{\left(-\frac{c}{2m}\right)^2 + \frac{k}{m} - \left(\frac{c}{2m}\right)^2} = \sqrt{\frac{k}{m}} \tag{7.36}$$

and:

$$\zeta = \frac{-\sigma}{\sqrt{\sigma^2 + \omega^2}} = \frac{\frac{c}{2m}}{\sqrt{\frac{k}{m}}} = \frac{c}{2\sqrt{km}} \tag{7.37}$$

and:

$$\tau = -\frac{1}{\sigma} = \frac{2m}{c} \tag{7.38}$$

Undamped Linear Second Order ODEs

When damping is present in a second order problem, the equations must be reduced to first order in order to cast it as an eigenvalue problem, but for undamped second order problems, it is possible to solve directly without reduction to first order. Consider the case where $\mathbf{L} = \mathbf{0}$.

$$\mathbf{M}\ddot{\boldsymbol{z}}(t) + \mathbf{K}\boldsymbol{z}(t) = \mathbf{0} \tag{7.39}$$

Allow that the solution is exponential:

$$\boldsymbol{z}(t) = \boldsymbol{z}_0 \mathrm{e}^{st} \tag{7.40}$$

Substitution into the equations of motion gives:

$$\mathbf{M}\boldsymbol{z}_0 s^2 \mathrm{e}^{st} + \mathbf{K}\boldsymbol{z}_0 \mathrm{e}^{st} = \mathbf{0} \tag{7.41}$$

or:

$$[\mathbf{M}s^2 + \mathbf{K}]\boldsymbol{z}_0 e^{st} = \mathbf{0} \tag{7.42}$$

A similar result to the first order system is obtained.

$$[\mathbf{M}s^2 + \mathbf{K}]\boldsymbol{z}_0 = \mathbf{0} \tag{7.43}$$

A multiplication by the inverse of **M** shows that the equations have been transformed into an eigenvalue problem.

$$[\mathbf{I}s^2 + \mathbf{M}^{-1}\mathbf{K}]\boldsymbol{z}_0 = \mathbf{0} \tag{7.44}$$

In this case, the eigenvalues of $-\mathbf{M}^{-1}\mathbf{K}$ will give the values of s^2, so the square root of the eigenvalue now gives s. In the case of two second order equations, there will be two eigenvalues, typically negative, such that their square roots will give both positive and negative imaginary values, four in total, indicative of an oscillatory solution. Each imaginary pair is one natural frequency, identical to the result from the first order approach.

This is in contrast to the case where an undamped second order vibration problem is reduced to first order. Here, the eigenvalues of **A** will directly give the frequencies of oscillation as imaginary roots, which will appear as conjugate pairs. For example, a system of two second order equations will become four first order equations, with four eigenvalues, appearing as two pairs of conjugates. Each pair of conjugates will represent one natural frequency, but no square root operation is required.

A further source of confusion is due to yet another possible solution technique that is sometimes encountered in the literature. The solution to an undamped linear second order ODE cannot be exponential, so the assumed solution may taken as a harmonic.

$$\boldsymbol{z}(t) = \boldsymbol{z}_0 \cos(\omega t) \tag{7.45}$$

In this case, the same eigenvalue problem will appear, but with a change in sign, i.e., the eigenvalues of $\mathbf{M}^{-1}\mathbf{K}$ are sought. Changing the sign of a matrix changes the sign of all the eigenvalues. As a result, in this convention, the eigenvalues will typically appear as positive, their square roots are the values of ω, and no imaginary numbers appear.

7.2.5 Frequency Response Analysis

Typically, when working with systems of first order equations, they will be formatted as follows:

$$\dot{\boldsymbol{x}}(t) = \mathbf{A}\boldsymbol{x}(t) + \mathbf{B}\boldsymbol{u}(t) \tag{7.46}$$

where **A** is the *system matrix* and **B** is the *input matrix*. The vector $\boldsymbol{x}(t)$ is referred to as the *state vector* and $\boldsymbol{u}(t)$ is the *input vector*. The equations are said to be cast in *state space* form. In this case, these equations are typically supplemented with another set of related equations:

$$\boldsymbol{y}(t) = \mathbf{C}\boldsymbol{x}(t) + \mathbf{D}\boldsymbol{u}(t) \tag{7.47}$$

The vector $\boldsymbol{y}(t)$ is referred to as the *output vector*, $\mathbf{C}$ is the *output matrix* and $\mathbf{D}$ is the *feedthrough* or *feedforward matrix*. The output matrix allows the calculation of additional information related to the states, e.g., the velocity of a specific point on a body, or the distance between two points on different bodies. The feedthrough matrix allows inputs to bypass the transient nature of the differential equation and act directly on the output. In many cases, the feedthrough will be zero, but one notable exception is in the case of a mechanical system where the output vector contains acceleration data. This form of the equations is frequently found in the study of feedback control theory. An equivalent but more compact notation is commonly used for the state space equations:

$$\begin{Bmatrix} \dot{\boldsymbol{x}} \\ \boldsymbol{y} \end{Bmatrix} = \begin{bmatrix} \mathbf{A} & \mathbf{B} \\ \mathbf{C} & \mathbf{D} \end{bmatrix} \begin{Bmatrix} \boldsymbol{x} \\ \boldsymbol{u} \end{Bmatrix} \tag{7.48}$$

Another useful result from linear analysis is the concept of *frequency response*. In the general case, a closed form for the particular solution resulting from the input function $\boldsymbol{u}(t)$ cannot be found. However, if the input to a linear system is strictly sinusoidal, then the forced output will also be sinusoidal, and at the same frequency as the forcing function. The two sinusoids will not necessarily be in phase with each other; the phase lag will be a function of the forcing frequency, as will the ratio of amplitudes. The problem can be solved by setting the both the input and the output to be sinusoids, but the phase lag results in a complicated expression that requires a significant amount of algebraic manipulation. As a more convenient mathematical alternative, a complex exponential form is used instead.

$$\boldsymbol{u}(t) = \boldsymbol{u}_0 e^{st},\ \boldsymbol{x}(t) = \boldsymbol{x}_0 e^{st},\ \boldsymbol{y}(t) = \boldsymbol{y}_0 e^{st} \tag{7.49}$$

Differentiating and substituting into the original equation gives:

$$\boldsymbol{x}_0 s e^{st} = \mathbf{A}\boldsymbol{x}_0 e^{st} + \mathbf{B}\boldsymbol{u}_0 e^{st} \tag{7.50}$$

$$\boldsymbol{x}_0 = [\mathbf{I}s - \mathbf{A}]^{-1}\mathbf{B}\boldsymbol{u}_0 \tag{7.51}$$

Now the output is:

$$\boldsymbol{y}(t) = \boldsymbol{y}_0 e^{st} = \mathbf{C}\boldsymbol{x}_0 e^{st} + \mathbf{D}\boldsymbol{u}_0 e^{st} \tag{7.52}$$

Substituting gives:

$$\boldsymbol{y}_0 = [\mathbf{C}[\mathbf{I}s - \mathbf{A}]^{-1}\mathbf{B} + \mathbf{D}]\boldsymbol{u}_0 \tag{7.53}$$

This result shows another important feature of LTI problems. The forced solution is linear in the inputs, i.e., if the amplitude of the input is scaled by some value, the forced solution scales by the same value. The matrix function $\mathbf{G}(s)$ is called the *transfer function* of the differential equation.

$$\mathbf{G}(s) = [\mathbf{C}[\mathbf{I}s - \mathbf{A}]^{-1}\mathbf{B} + \mathbf{D}] \tag{7.54}$$

The frequency response problem is solved by setting the variable $s = \mathrm{i}\omega$, where ω is varied over a range of frequencies of interest, with the result representing the sensitivity of the system to frequencies in this range. The astute reader may notice that the same result is achieved by considering the Laplace transform of the ODE, other than the terms following from the initial conditions. The omission of the initial condition terms is consistent with the intention to find only the forced solution of the ODE. It is assumed that the initial conditions influence the free solution only, as found from the eigenvalue problem, and that the free solution decays after some time, leaving only the forced solution.

One may question the validity of selecting a complex forcing function $\boldsymbol{u}(t) = \boldsymbol{u}_0 \mathrm{e}^{\mathrm{i}\omega t}$, if the equations are meant to represent a physical system. However, this may be done safely, due to the principle of linear superposition, which states:

$$f(a + b) = f(a) + f(b) \tag{7.55}$$

The response of a linear system to a sum of inputs is the same as the sum of the responses to the individual inputs.

$$f(2a) = 2f(a) \tag{7.56}$$

More importantly, for the complex case:

$$f(a + \mathrm{i}b) = f(a) + \mathrm{i}f(b) \tag{7.57}$$

The response of a linear system to a complex input is also complex, but there is no ‘crossover’ in the system, i.e., the real part of the response is the response to the real part of the input, and the imaginary part of the response is the response to the imaginary part of the input. Real inputs do not create imaginary responses, and imaginary inputs do not create real responses. The complex exponential is simply a convenient method to generate both the real and imaginary solutions, and the imaginary component of the solution may be safely ignored.

The resulting calculation of $\mathbf{G}(\mathrm{i}\omega)$ will generate a matrix of complex numbers representing the sensitivity of the model, i.e., the ratio of output to input. The temptation may be to discard the imaginary part of this matrix, but this is incorrect. It is important to recall that the imaginary part of the sensitivity can still generate a real component of the solution. The imaginary part of the sensitivity simply represents the out-of-phase part of the response. Consider a single input single output system, where $\mathbf{G}(\mathrm{i}\omega)$ collapses to a scalar value, say $a + b\mathrm{i}$, and $\boldsymbol{u}_0 = 1$:

$$\begin{aligned} y(t) &= (a + b\mathrm{i})(1)\mathrm{e}^{\mathrm{i}\omega t} = (a + b\mathrm{i})(\cos\omega t + \mathrm{i}\sin\omega t) \\ &= \underbrace{a\cos\omega t - b\sin\omega t}_{\text{real part}} + \underbrace{a\mathrm{i}\sin\omega t + b\mathrm{i}\cos\omega t}_{\text{imaginary part}} \end{aligned} \tag{7.58}$$

The two real sinusoidal terms represent the solution to the real input, and the two imaginary sinusoids may be discarded. The magnitude of the resulting sum of the two real sinusoids is the root of the sum of the squares of their amplitudes, and the phase angle between the input and output of the system depends on their relative amplitudes. The values of $\mathbf{G}(\mathrm{i}\omega)$ are typically presented as the magnitude and phase of the complex number, rather than as the real and imaginary terms.

Consider once again a single second order equation, now with a forcing function:

$$m\ddot{z}(t) + c\dot{z}(t) + kz(t) = f(t) \tag{7.59}$$

Reducing to first order gives the system and input matrices:

$$\begin{Bmatrix} \dot{z}(t) \\ \ddot{z}(t) \end{Bmatrix} = \begin{bmatrix} 0 & 1 \\ -k/m & -c/m \end{bmatrix} \begin{Bmatrix} z(t) \\ \dot{z}(t) \end{Bmatrix} + \begin{bmatrix} 0 \\ 1/m \end{bmatrix} \{f(t)\} \tag{7.60}$$

Adding the output equation gives the output and feedthrough matrices:

$$z(t) = \begin{bmatrix} 1 & 0 \end{bmatrix} \begin{Bmatrix} z(t) \\ \dot{z}(t) \end{Bmatrix} + [0]f(t) \tag{7.61}$$

The system matrices can now be used to find the transfer function:

$$\begin{aligned} \mathbf{G}(s) &= \begin{bmatrix} 1 & 0 \end{bmatrix} \left[\begin{bmatrix} s & 0 \\ 0 & s \end{bmatrix} - \begin{bmatrix} 0 & 1 \\ -k/m & -c/m \end{bmatrix} \right]^{-1} \begin{bmatrix} 0 \\ 1/m \end{bmatrix} + [0] \\ &= \begin{bmatrix} 1 & 0 \end{bmatrix} \begin{bmatrix} s & -1 \\ k/m & s + c/m \end{bmatrix}^{-1} \begin{bmatrix} 0 \\ 1/m \end{bmatrix} \\ &= \frac{\begin{bmatrix} 1 & 0 \end{bmatrix} \begin{bmatrix} s + c/m & 1 \\ -k/m & s \end{bmatrix} \begin{bmatrix} 0 \\ 1/m \end{bmatrix}}{s^2 + (c/m)s + k/m} \\ &= \frac{1/m}{s^2 + (c/m)s + k/m} \\ &= \frac{1}{ms^2 + cs + k} \end{aligned} \tag{7.62}$$

The expression $s = \mathrm{i}\omega$ is substituted into the transfer function, effectively enforcing the assumption that $f(t) = f_0 \sin\omega t$ and $x(t) = x_0 \sin(\omega t + \phi)$.

$$\mathbf{G}(\mathrm{i}\omega) = \frac{1}{k - m\omega^2 + c\omega\mathrm{i}} \tag{7.63}$$

For real values of the parameters, the frequency response function will be a complex number. Its magnitude and phase can be computed. Recall that the magnitude of a fraction when both numerator and denominator are complex can be found from:

$$\left|\frac{a+bi}{c+di}\right| = \frac{\sqrt{a^2+b^2}}{\sqrt{c^2+d^2}} \tag{7.64}$$

and the phase is:

$$\angle\left(\frac{a+bi}{c+di}\right) = \tan^{-1}\left(\frac{b}{a}\right) - \tan^{-1}\left(\frac{d}{c}\right) \tag{7.65}$$

So the magnitude of the transfer function is:

$$|\mathbf{G}(\mathrm{i}\omega)| = \frac{x_0}{f_0} = \frac{1}{\sqrt{(k-m\omega^2)^2+(c\omega)^2}} = \frac{1}{m\sqrt{(\omega_\mathrm{n}^2-\omega^2)^2+(2\zeta\omega_\mathrm{n}\omega)^2}} \tag{7.66}$$

and the angle:

$$\angle\mathbf{G}(\mathrm{i}\omega) = \phi = \tan^{-1}\left(\frac{0}{1}\right) - \tan^{-1}\left(\frac{c\omega}{k-m\omega^2}\right) = -\tan^{-1}\left(\frac{2\zeta\omega_\mathrm{n}\omega}{\omega_\mathrm{n}^2-\omega^2}\right) \tag{7.67}$$

An examination of the result will reveal several important points. First, if the frequency ω is taken to zero, the mass and damping terms will have no effect and the magnitude will depend only on the stiffness. Also, the phase lag will go to zero. If the frequency is taken to a very large value, the magnitude tends toward zero, and phase lag tends to π radians. However, in the intermediate range of frequencies, if the damping $\zeta < 1/\sqrt{2}$, then the amplitude of motion can increase above the steady state value. In the region near the natural frequency, the amplitudes can become very large, and the smaller the value of ζ, the larger the amplitude becomes. This phenomenon is known as *resonance*. The peak amplitude occurs when the excitation frequency ω equals the resonant frequency ω_r, where:

$$\omega_\mathrm{r} = \sqrt{1-2\zeta^2}\omega_\mathrm{n} \tag{7.68}$$

Note that the resonant frequency is distinct from both the natural and damped natural frequencies, but in cases where the damping is small enough to allow a significant resonance effect, all three frequencies will be very close together. At the resonant frequency, the amplitude can be found by substituting $\omega = \omega_\mathrm{r}$, giving:

$$|\mathbf{G}(\mathrm{i}\omega_\mathrm{r})| = \frac{x_0}{f_0} = \frac{1}{2k\zeta\sqrt{1-\zeta^2}} \tag{7.69}$$

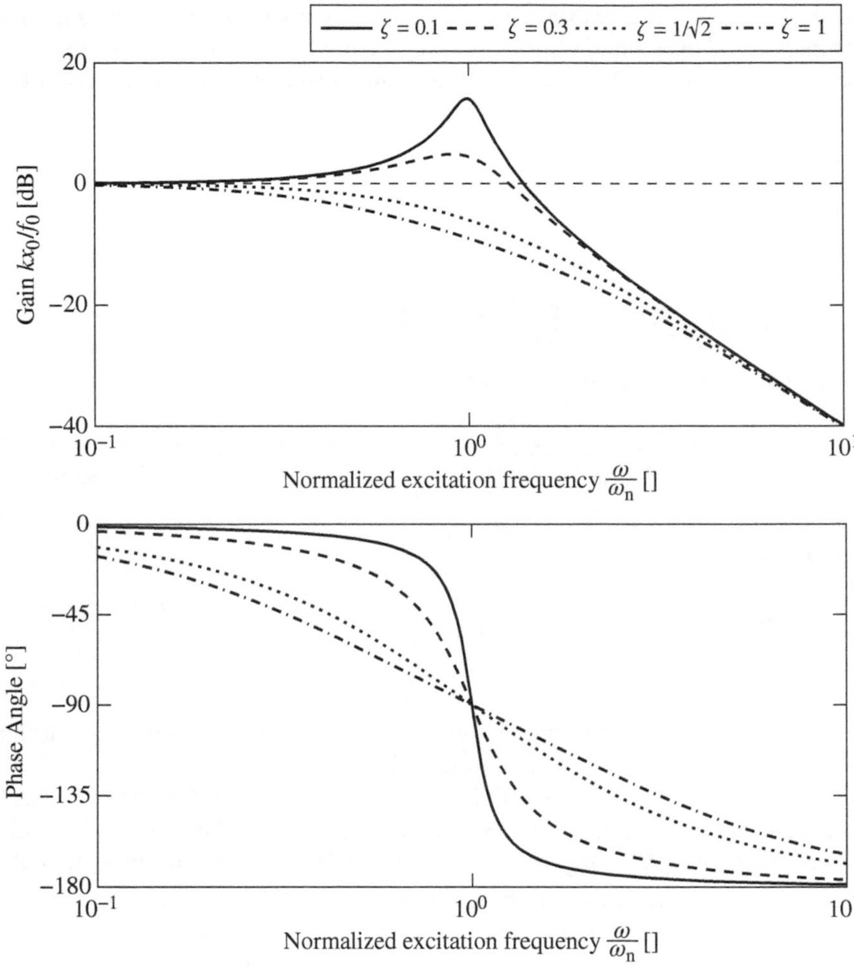

Figure 7.2 The motion of a spring mass damper in response to an applied sinusoidal force and the resulting phase lag are both functions of frequency. The lightly damped case shows a resonance near the natural frequency. The system has 90° of phase lag at the natural frequency, for any damping. Cases where the damping ratio $\zeta > 1/\sqrt{2}$ will prevent resonance.

Note also that when the excitation frequency equals the natural frequency, $\omega = \omega_n$, that the phase lag $\phi = \pi/2$. Due to proximity of the natural and resonance frequencies, at resonance the phase angle will typically remain close to $\pi/2$.

Plots of the magnitude and phase are shown in Figure 7.2. Note that the plots are displayed using a logarithmic scale on the abscissa for frequency, and the magnitude is plotted as the *gain*, using units of decibels (dB) on the ordinate axis. This is a common practice for frequency response, as the curves have asymptotes when plotted in this fashion. With this convention, each increment of 20 dB in the gain represents a factor of ten in the magnitude, i.e., 0 dB is a ratio of 1:1; output amplitude equals input amplitude. A 20 dB gain means a 10:1 ratio in output amplitude to input amplitude; a gain below −40 dB means the output amplitude is less than 1% of the input amplitude.

7.3 Differential Algebraic Equations

The final classification of mathematical problems is the differential algebraic equation (DAE). As the name implies, these sets contain both differential and algebraic equations that are coupled to each other. Unfortunately, differential algebraic equations are often particularly challenging to solve, and they occur frequently in the equations of motion of mechanical systems. Consider the definition of an ODE, as in Equation (7.5), reproduced here:

$$\boldsymbol{f}(\boldsymbol{x}(t), \dot{\boldsymbol{x}}(t), t) = \boldsymbol{0} \tag{7.70}$$

where the condition:

$$\det \left[\frac{\partial \boldsymbol{f}}{\partial \dot{\boldsymbol{x}}} \right] = 0 \tag{7.71}$$

applies. This condition implies that the $\dot{\boldsymbol{x}}$ terms cannot be isolated, and so no explicit form exists. Some of the associated expressions then reduce to algebraic equations. An example first order DAE system could look like:

$$\begin{Bmatrix} \dot{x}_1 + \dot{x}_2 - 3x_1 - 4x_2 - 3x_3 - t \\ \dot{x}_1 - \dot{x}_2 - 3x_1 - 4x_2 + 3x_3 + t \\ \dot{x}_1 - 2x_1 - 4x_2 + x_3 - 1 \end{Bmatrix} = \begin{Bmatrix} 0 \\ 0 \\ 0 \end{Bmatrix} \tag{7.72}$$

A check of the condition in Equation (7.71) will confirm that the example is indeed a DAE. A closer examination shows that the first two equations will quickly allow isolation of $\dot{x}_1$ and $\dot{x}_2$. However, the third equation then collapses, and the result becomes:

$$\begin{Bmatrix} \dot{x}_1 \\ \dot{x}_2 \\ 0 \end{Bmatrix} = \begin{Bmatrix} 3x_1 + 4x_2 \\ 3x_3 + t \\ x_1 + x_3 - 1 \end{Bmatrix} \tag{7.73}$$

In general, the strategy that is typically used in the numerical solution of DAEs is to select either a differential or an algebraic equation approach, and to convert those equations that are of the opposing form, so that the problem is expressed entirely as a set of algebraic equations, or entirely as a set of differential equations. An unexpected consequence of the differential equation approach is that while it will almost always produce a complete solution, that solution may contain significant numerical error. In contrast, the algebraic equation approach tends to have much better resistance to numerical error, but occasionally it fails to converge, and produces an incomplete solution. The analyst may be left in the unfortunate position of choosing between a clearly flawed solution or no solution at all.

7.3.1 Differential Equation Approach

The first approach is to replace the algebraic equations with differentiated versions of themselves. In the example above, the third equation can be differentiated to find an expression for $\dot{x}_3$, to give a set of three ODEs. This effectively converts the problem from a DAE to an ODE that can be solved by any of the previously mentioned ODE algorithms. The example problem becomes:

$$\begin{Bmatrix} \dot{x}_1 \\ \dot{x}_2 \\ \dot{x}_3 \end{Bmatrix} = \begin{Bmatrix} 3x_1 + 4x_2 \\ 3x_3 + t \\ -3x_1 - 4x_2 \end{Bmatrix} \tag{7.74}$$

If one selects Euler's method:

$$\boldsymbol{x}_{n+1} = \boldsymbol{x}_n + hf(\boldsymbol{x}_n, t_n) = \boldsymbol{x}_n + h\dot{\boldsymbol{x}}_n \tag{7.75}$$

In the case of the example problem, if one defines:

$$\boldsymbol{x}_{n+1} = \begin{Bmatrix} x'_1 \\ x'_2 \\ x'_3 \end{Bmatrix} \tag{7.76}$$

After substitution of the complete ODE set into Euler's method, the solution is given by repeated application of:

$$\begin{Bmatrix} x'_1 \\ x'_2 \\ x'_3 \end{Bmatrix} = \begin{Bmatrix} (1 + 3h)x_1 + 4hx_2 \\ x_2 + 3hx_3 + ht \\ -3hx_1 - 4hx_2 + x_3 \end{Bmatrix} \tag{7.77}$$

Note that a subtle change occurs when this approach is used. Enforcing the differentiated form of the algebraic equation does not necessarily enforce the original form, as a constant offset can be introduced by the integration. If the algebraic equations are required to be differentiated more than once in order to build the set of ODEs, the situation is even worse. Unfortunately, as a result, this approach can sometimes produce numerical round-off errors that compound until original algebraic equations are clearly violated.

7.3.2 Algebraic Equation Approach

The second approach is to substitute an ODE solver into the differential equations in the set. This effectively converts the problem to an algebraic only problem, which can then be solved using a Newton–Raphson or some other similar solver. Again, if one selects Euler's method, and solves for the rate of change:

$$\dot{\boldsymbol{x}}_n = \frac{\boldsymbol{x}_{n+1} - \boldsymbol{x}_n}{h} \tag{7.78}$$

The ODE solver can be used to find expressions to replace $\dot{x}_1$ and $\dot{x}_2$, while the original algebraic equation is used directly.

$$\begin{Bmatrix} \frac{1}{h}(x_1' - x_1) - 3x_1 - 4x_2 \\ \frac{1}{h}(x_2' - x_2) - 3x_3 - t \\ x_1' + x_3' - 1 \end{Bmatrix} = \begin{Bmatrix} 0 \\ 0 \\ 0 \end{Bmatrix} \tag{7.79}$$

$$\begin{bmatrix} \frac{1}{h} & 0 & 0 \\ 0 & \frac{1}{h} & 0 \\ 1 & 0 & 1 \end{bmatrix} \begin{Bmatrix} x_1' \\ x_2' \\ x_3' \end{Bmatrix} = \begin{Bmatrix} \frac{1}{h}x_1 + 3x_1 + 4x_2 \\ \frac{1}{h}x_2 + 3x_3 + t \\ 1 \end{Bmatrix} \tag{7.80}$$

The result is a set of algebraic equations that can be solved for x_1', x_2', and x_3'. In this example, the resulting set of algebraic equations is linear, because the original DAE was linear, but frequently the resulting algebraic equations will be nonlinear, and an iterative solution is required. In practice, a more sophisticated algorithm than Euler's method would be substituted, as there are many that offer improved performance, but the concept is the same.

This second approach can also suffer from numerical problems during the solution. It is often the case where some terms will have the stepsize h appear in the denominator. Very small values of h will then produce very large values in the expressions, which when added to the remaining much smaller values, will tend to introduce numerical round-off error. The unfortunate case can occur where the value of h is too large to allow the solution to proceed accurately, but any smaller value results in a poorly conditioned system of equations that cannot be solved. In this situation, the algebraic equation approach may simply fail to converge, and stall after only a partial solution. Nevertheless, this is the approach favoured by many of the large commercial multibody dynamics software tools.

7.3.3 Linear Differential Algebraic Equations

A third possibility exists that avoids the need to employ either of the techniques above. In the example, one may simply solve the third equation for x_3, and eliminate it from the first two equations. The set is reduced in dimension by one, and can be solved as an ODE. Once the complete solution for x_1 is found, then x_3 can be computed. The example problem becomes:

$$\begin{Bmatrix} \dot{x}_1 \\ \dot{x}_2 \end{Bmatrix} = \begin{Bmatrix} 3x_1 + 4x_2 \\ -9x_1 - 12x_2 + t \end{Bmatrix} \tag{7.81}$$

Unfortunately, in the general case, it may not always be possible to isolate the appropriate terms in order to eliminate them. However, if both the differential and algebraic equations in the set are linear, this option is always available, and is generally preferred. Linear equations are often written in *descriptor state space form.*

$$\begin{bmatrix} \mathbf{E} & \mathbf{0} \\ \mathbf{0} & \mathbf{I} \end{bmatrix} \begin{Bmatrix} \dot{x} \\ y \end{Bmatrix} = \begin{bmatrix} \mathbf{A} & \mathbf{B} \\ \mathbf{C} & \mathbf{D} \end{bmatrix} \begin{Bmatrix} x \\ u \end{Bmatrix} \tag{7.82}$$

When written in this form, it is the condition of the **E** matrix that determines if the system is a DAE or an ODE. If the **E** matrix can be inverted, the system is a set of ODEs, and can be easily converted to standard state space form. If the **E** matrix is singular, then equations are a set of DAEs. In this case, the conversion to standard form is still possible, but more complicated. It typically involves a technique known as *singular value decomposition* (SVD) to isolate and remove the algebraic terms, and will arrive at a standard state space that is smaller in dimension than the descriptor form.

Appendix A

Numerical Methods

A.1 Algebraic Equations

There are many methods for the solution of nonlinear algebraic equations that depend on *recursion*. A recursive method is an algorithm that will be applied repeatedly to its own result until the solution is reached. The most popular of these is the Newton–Raphson method. The process begins with an initial estimate for the solution $\boldsymbol{x}_1$. The estimate is substituted into the system of equations, and the difference from zero, or the *residual*, is found. The residual is then used to compute an improved estimate for the solution. The improved estimate is then used in turn to recompute the residual, and the process repeats. The method requires the *Jacobian* matrix of the function $\boldsymbol{f}$, i.e., the partial derivative of each function in the system with respect to each of the dependent variables, or in the case of the closely related secant method, an estimate of the Jacobian found using finite differences. The Newton–Raphson method can be written as:

$$\boldsymbol{x}_{n+1} = \boldsymbol{x}_n - \left.\frac{\partial \boldsymbol{f}}{\partial \boldsymbol{x}}\right|_{\boldsymbol{x}_n}^{-1} \boldsymbol{f}(\boldsymbol{x}_n, t) \tag{A.1}$$

where the subscript n refers to the iteration number. Once the estimate passes some sort of check, e.g., it is no longer changing significantly with further iteration, or the residual is sufficiently small, it is taken as the solution.

Note that when dealing with dynamic systems that the solution will be a function of time. The method works by *discretization* of the solution. During the iterative process, the value of t is held fixed. Rather than having a solution for $\boldsymbol{x}$ as a continuous function of t, only a discrete set of points in time are evaluated. To progress the solution over a given timespan, the value of t is incremented, and the recursive process is repeated for each new intermediate value. As the value of t typically changes by a small amount, the previous solution of $\boldsymbol{x}$ is usually a good candidate for the initial estimate at the next point in time. Note also that in the expression above, the inverse of the Jacobian matrix is used

Fundamentals of Vehicle Dynamics and Modelling: A Textbook for Engineers with Illustrations and Examples, First Edition. Bruce P. Minaker.

Companion website: www.wiley.com/go/minaker/vehicle-dynamics

to show the relationship, but this inverse is rarely computed in practice, as matrix inversion is computationally expensive and introduces numerical error. Instead, the product of the inverse and the function will be computed directly.

One might expect that because the Newton–Raphson method is suitable for the more general case of nonlinear equations, it would be equally effective when applied to linear systems. However, this is not the case; it is of no benefit to the solution of linear systems of equations, as it effectively solves a nonlinear problem by approximating it as a sequence of linear problems. It is the converse of the expectation that is true; the techniques used for the efficient solution of linear systems may be applied at each iteration to improve the Newton–Raphson algorithm.

The primary concern with this approach is the same with most recursive methods: a chance that the method will not converge, or will converge to a solution other than the one being sought. In this case, one possibility is to use *relaxation*, i.e., rather than use the new estimate as computed, a weighted average of the new estimate and the previous estimate is used.

$$\begin{aligned} \boldsymbol{x}_{n+1} &= \alpha \left(\boldsymbol{x}_n - \left.\frac{\partial \boldsymbol{f}}{\partial \boldsymbol{x}}\right|_{\boldsymbol{x}_n}^{-1} \boldsymbol{f}(\boldsymbol{x}_n, t) \right) + (1-\alpha)\boldsymbol{x}_n \\ &= \boldsymbol{x}_n - \alpha \left.\frac{\partial \boldsymbol{f}}{\partial \boldsymbol{x}}\right|_{\boldsymbol{x}_n}^{-1} \boldsymbol{f}(\boldsymbol{x}_n, t) \end{aligned} \tag{A.2}$$

Sometimes varying the relaxation parameter α can improve the convergence properties, but there is still no guarantee. Other similar methods for speeding or slowing the convergence are also available. Despite the convergence issue, this type of approach has been used successfully by a number of researchers and programmers.

A.2 Differential Equations

There are are many algorithms capable of providing a *numerical* solution to a differential equation, but the two most common families are classified as *linear multi-step* (LMS) methods, or *Runge–Kutta* (RK) methods. Both approaches have been widely successfully employed, and neither is clearly superior for all types of problems, although RK methods are generally the best choice for a first attempt when facing a new problem. Both types of methods require discretization in time, and solutions are given at only a finite number of points. Both the LMS and RK families have examples that are *implicit* and require a recursive solution method at each time step; likewise there are also *explicit* methods in each family.

The primary distinction is that LMS methods use information from previous solutions in time in order to predict the solution at the next time step, where

RK methods ignore the past, and instead use multiple tentative steps to gather information before stepping the solution forward in time.

Multi-Step Methods

The generic LMS method can be written as:

$$\sum_{i=0}^{k} a_i \boldsymbol{x}_{n+i} = h \sum_{i=0}^{k} b_i \boldsymbol{f}(\boldsymbol{x}_{n+i}, t_{n+i}) \tag{A.3}$$

The choices of k, a_i, and b_i determine the specific method in the family. For example, setting $k = 1, a_0 = -1, a_1 = 1, b_0 = -1, b_1 = 0$, gives Euler's method, the simplest numerical method for solving an ODE.

$$\boldsymbol{x}_{n+1} = \boldsymbol{x}_n + h\boldsymbol{f}(\boldsymbol{x}_n, t_n) \tag{A.4}$$

The subscript n again refers to the iteration number, but unlike the Newton–Raphson algorithm, with each application of Euler's method, time is incremented by the stepsize h. The method progresses by approximating the curved true solution by a straight line of equal slope at the point $(\boldsymbol{x}_n, t_n)$. This slope is multiplied by the time step h to find the change in the dependent variable, which is then added to the initial value. Of course, Euler's method does not give the true solution, unless the true solution also happens to be a straight line, but if the step size h is chosen sufficiently small, then the numerical solution will be a good approximation to the true solution. While shrinking the step size will force the numerical solution to converge to the true solution, it also increases the number of steps required, and therefore the computational effort. The optimal step size to balance accuracy and effort is problem-dependent, and often requires some experimentation to find a suitable value. While illustrative, Euler's method is rarely used in practice, as there are many other methods that provide a better trade-off between accuracy and effort.

The multi-step methods generally fall into one of three important categories: Adams–Bashforth, Adams–Moulton, and Gear, sometimes referred to as Backwards Differentiation Formulae (BDF). The Adams-Bashforth methods are explicit methods, meaning that the desired unknown $\boldsymbol{x}_{n+1}$ can be isolated and solved. The Adams–Moulton and Gear methods are implicit, meaning that the solution for $\boldsymbol{x}_{n+1}$ depends on $f(\boldsymbol{x}_{n+1})$, which in turn depends on $\boldsymbol{x}_{n+1}$, so a recursion is required at each step to advance the solution. The recursive methods require an initial guess to begin, but fortunately, the solution from the previous time step will typically provide a good starting point.

Both Adams methods are characterized by only two nonzero values of a_i, i.e., the slope of the solution at many points may be required, but the value itself is only used at two points; of these two, one value is known, and the other is sought. Because only two points are used, the Adams methods can be considered to use a calculation of the 'effective' slope between the two points. The

Gear methods are similar but opposite; there are only two nonzero values of b_i. Only two values of the slope of the solution are required, but the value of the solution at many points is used. These points are typically known from the past values of the solution.

Runge–Kutta Methods

The RK methods are different than the multi-step methods, in that they never look to the past for information. Instead, they extend 'feelers', or tentative steps forward, to gain information about the future. The information from these tentative steps is gathered and averaged, until the actual step is taken. One benefit of ignoring past information is that simulations using RK methods can be started conveniently. At the first step, when no past information is available, LMS methods may require a complicated procedure to estimate prior solution values in order to initiate the solution. The general RK method can be written as:

$$\boldsymbol{x}_{n+1} = \boldsymbol{x}_n + h \sum_{i=1}^{s} b_i \boldsymbol{k}_i \tag{A.5}$$

where:

$$\boldsymbol{k}_i = \boldsymbol{f}(\boldsymbol{x}_n + h \sum_{j=1}^{s} a_{ij} \boldsymbol{k}_j, t_n + c_i h) \tag{A.6}$$

Similarly to the LMS methods, the values of a_{ij}, b_i, and c_i determine the specific method in the family. For example, setting $s = 1$, $a_{11} = 0$, $b_1 = 1$, and $c_1 = 0$ gives Euler's method. The definitions of LMS and RK methods overlap in this one instance.

One of the more popular RK methods is a four stage method that was very widely used. In fact, the four stage method was so popular that many students are surprised to learn that the RK methods are an entire family, as the four stage method is often introduced as *the* RK method. More recently, a five stage method, known as the Dormand and Prince or DOPRI54 method, has gained widespread popularity for its good performance over a wide range of problems.

The RK methods have a few interesting features that make them very popular. First, it is relatively easy to develop *embedded* methods, where two solutions can be found simultaneously, with the second of higher order, and little additional computational cost. The difference in the two solutions can then be used to estimate the error, and and in turn, adjust the size of the next step. A variable step size algorithm is much easier to implement in RK methods than in LMS methods, due to the forward step only nature of the method. Second, some RK methods have been developed using a *first-same-as-last* (FSAL) strategy, where the first stage of a step can be avoided by reusing the calculation from the last stage of the previous step, improving the overall computational efficiency. Finally, some RK methods known as *continuous explicit* (CERK) methods, have been developed such that the solution provided is not the value of the function at the end of the step, but rather the coefficients of a polynomial that fit the solution over the entire step. This can greatly improve the efficiency if the solution is required to be presented at a smaller time interval than is necessary for calculation accuracy, e.g., for plotting. In practice, however, the bottleneck is usually the calculation stepsize rather than the plot stepsize, and in this case the CERK methods offer no substantial performance advantage.

A factor that can often greatly influence the success of a particular method is how it responds to *stiff* problems. The term does not refer to the property of mechanical stiffness, but rather to the equations that govern the behaviour of a system. There are many formal definitions of stiffness, but generally it refers to problems where two or more vastly different time scales are required to capture all the relevant behaviour, e.g., a vibrating system with two very different natural frequencies. In these types of problems, algorithms such as the DOPRI54 that usually give good general performance can fail quite dramatically, with a several hundredfold increase in the number of steps required and, correspondingly, the computational time. As a general rule, implicit methods tend to be relatively immune to the effects of stiffness, and are often required in order to have any practical solution to a stiff problem.

Linear Time History Solutions

Finally, if the time history solution of a linear system is desired, it is possible to avoid both the LMS or RK methods and instead make use of the *matrix exponential*. The standard numerical solvers will function and give a proper solution, but they are not as computationally efficient. The matrix exponential allows time history solutions to be found very quickly when compared to the LMS or RK methods. The method assumes that the solution is computed in discrete time, using a fixed time step T only, but the eigenvalues give the necessary information to ensure that the step size can be chosen optimally in advance, rather than being continually adjusted as the solution progresses. The process begins by converting the differential equation to a *difference equation*. The equivalent difference equation is:

$$\boldsymbol{x}_{n+1} = \mathbf{A}_\mathrm{d}\boldsymbol{x}_n + \mathbf{B}_\mathrm{d}\boldsymbol{u}_n \tag{A.7}$$

where $\boldsymbol{x}_n = \boldsymbol{x}(nT)$. The solution to the difference equation can be computed sequentially, given $\boldsymbol{x}_0$ and $\boldsymbol{u}_n$. There is some additional initial computational cost for the conversion to the difference equation, but once done, the solution of the difference equation is very efficient. The conversion relies on the matrix exponential, which is written as:

$$\boldsymbol{\Phi}(t) = \mathrm{e}^{\mathbf{A}t} = \sum_{k=0}^{\infty} \frac{1}{k!}\mathbf{A}^k t^k \tag{A.8}$$

The name and notation used for the matrix exponential is due to the similarity to the Taylor series of the scalar exponential function. The discrete time terms are found as:

$$\mathbf{A}_{\mathrm{d}} = \boldsymbol{\Phi}(T) \approx \mathbf{I} + T\mathbf{A} \tag{A.9}$$

$$\mathbf{B}_{\mathrm{d}} = \mathbf{A}^{-1}(\boldsymbol{\Phi}(T) - \mathbf{I})\mathbf{B} \approx T\mathbf{B} \tag{A.10}$$

The efficient and accurate calculation of the matrix exponential can be challenging, and typically relies on some type of recursion that terminates after a finite number of terms are summed. Approximations for $\mathbf{A}_{\mathrm{d}}$ and $\mathbf{B}_{\mathrm{d}}$ can be found relatively easily and quickly, but careful inspection reveals that if these are used, the result is equivalent to the application of Euler's method to the problem, which is not as computationally efficient as some of the other options.

Index

Fundamentals of Vehicle Dynamics and Modelling: A Textbook for Engineers with Illustrations and Examples,
First Edition. Bruce P. Minaker.

Companion website: www.wiley.com/go/minaker/vehicle-dynamics